HIGH TECHNOLOGY MARKETING MANAGEMENT

WITHDRAWN

HIGH TECHNOLOGY MARKETING MANAGEMENT

Robert A. Rexroad

A Ronald Press Publication
JOHN WILEY & SONS
New York   Chichester   Brisbane   Toronto   Singapore

Copyright © 1983 by John Wiley & Sons, Inc.

All rights reserved. Published simultaneously in Canada.

Reproduction or translation of any part of this work beyond that permitted by Section 107 or 108 of the 1976 United States Copyright Act without the permission of the copyright owner is unlawful. Requests for permission or further information should be addressed to the Permissions Department, John Wiley & Sons, Inc.

Library of Congress Cataloging in Publication Data:

Rexroad, Robert A., 1924–
   High technology marketing management.

  "A Ronald Press publication."
  Includes index.
  1. Marketing management—United States. 2. Technology—United States—Marketing. 3. Government purchasing—United States. I. Title. II. Series.

HF 5415.13.R44 1983 658.8 83-3448
ISBN 0-471-86877-9

Printed in the United States of America

10 9 8 7 6 5 4 3 2 1

PREFACE

I envy the many newcomers to the marketing profession who have selected the high technology industry as the launching platform for a very exciting and rewarding career. The growth pattern should be phenomenal for almost all segments of high technology industry. Hundreds of new industries will be created; thousands of new and exciting products will be developed; and certainly the marketer and the marketing manager will play a vital role in filling the needs of the marketplace, thus contributing to company success as well as personal growth.

This book addresses the marketing management function in dealing with high technology products where the government is the direct or indirect customer. It is an extremely complex field where the disciplines, and management responses required, are often dictated by the largest customer—the government.

Government marketing and its management have not been given comprehensive treatment previously because many of the techniques and controls employed by the participating companies are not widely discussed. Most companies do not reveal their specific practices or their win strategy formulations for proprietary reasons. The key contributing factors to this are the competitive nature of the marketplace and the process of dealing with classified government program information.

While I obviously address high technology, and primarily the government and industry marketplace, many of the management disciplines discussed will find application wherever the complexities of company–customer–competitor need to be better understood. That triad is commonplace throughout the business world; I simply bring this relationship to the forefront, to serve as the basic structure of a systems approach to the marketing (sales) management function.

My gratitude extends to many: to long-time industry and government associates; to an exceptional secretary, Susan Knake, who typed a readable manuscript from unkempt drafts; to the publisher's patient and skillful editorial staff; and to my wife, Helen, who believed the task would be completed and then made it possible in a number of ways.

A final note: In order to provide a smooth flowing text, I avoided using the awkward sounding "he or she." Instead, I used the generic "he" when referring to both genders, simply for grammatical and writing purposes.

ROBERT A. REXROAD

Fort Wayne, Indiana
June 1983

CONTENTS

List of Figures   xiii

1. HIGH TECHNOLOGY MARKETING MANAGEMENT: AN OVERVIEW   1
   The Definition of High Technology 2
   Marketing and High Technology 3
   Marketing the Government Customer 5
   Marketing or Sales—or Both 7
   The Active Role in Marketing: The Systems Approach 9

2. COMPANY STRATEGY AND PLANNING   11
   Defining the Terms 12
   The Business Plan 13
   Strategy Implementation 17
   Planning Guidelines 17
   Forecasting 20
   Investment Strategy 22
   The Environment 23
   The Marketing Manager's Role in Business Planning 24
   A Guide for Business Plan Formulation 25

3. **VIEWING THE MARKETING FUNCTION AS A TOTAL SYSTEM** 27
   The Systems Objective 28
   Systems Diagram Construction and Use 30
   Value of Systems Approach 35

4. **THE HIGH TECHNOLOGY MARKETING ORGANIZATION** 39
   Organizational and Operational Relationships 40
   The Company Management Hierarchy and Work Performance 42
   Supplier Relationships 45
   Relationships Within the Company 46
   The Marketing Organization 48

5. **DEFINING THE COMPANY TECHNOLOGY BASE** 53
   Technology Base Evaluation 54
   New Business Targets and the Technology Base 56
   Maintenance of the Technology Base 58
   Technology Applications 59
   Technology Base Management 60

6. **THE CUSTOMER** 63
   Customer Listings 64
   Customer Listing Evaluations 66
   Customer Business Practices 68
   Customer Listing Expansion 70
   Managing the Customer Listing 70

7. **COMPETITOR ANALYSIS** 73
   Defining the Competitor and Compiling a Dossier 74
   Competitor Evaluation 77

8. HIGH TECHNOLOGY MARKET ANALYSIS   81
   A Definition   82
   The Total Market   83
   The Available Market   87
   The Served Market   87
   Market Analysis Values   88

9. TARGET SELECTION AND WIN STRATEGY   91
   The Company–Customer Relationship   92
   Selecting Targets   96
   Company Performance and Target Selection   100
   Win Strategy   101

10. FORECASTING   109
    Forecasting for Near-term Programs   110
    Forecasting for Long-range Programs   113
    The Flexibility of Forecasts   115

11. PROPOSALS AND PRESENTATIONS   117
    Implementation of Proposal Techniques   118
    Proposal Ranking   120
    Managing the Small Proposal   120
    Proposal Content   121
    The Cost Proposal   122
    Special Proposal Considerations   125

12. RISK MANAGEMENT   127
    Risk Definitions   128
    Risk Identification   129
    Managing Risk   129
    Overlooked Risk   131

13. CONTRACTS, PRICING,
    AND NEGOTIATION   133
    The Government Procurement
       (Acquisition) System   134

Contract Types 136
Pricing Strategies and Effects 140
The Buy-in 142
The Best and Final Offer 143
Terms and Conditions 145
Audit and Negotiation 146
Other Points to Consider 148
Improving Customer–Company Relationships 148

14. MANAGEMENT OF MARKETING ASSETS 151
    The R&D Account 152
    The B&P Account 154
    A Control System 155
    The Budget Process 156

15. CONTROL SYSTEMS 159
    A Case for a Control System 159
    The Basic Controls 161

16. MARKETING DATA ACQUISITION AND CONTROL 165
    A Guide for Data Acquisition 166
    The Data Service Organization 168
    Classified Data 168
    Beginning a Data System 169

17. A MARKETING MANAGEMENT INFORMATION SYSTEM 173
    A Concept for Information Management 174

18. SECURITY 179
    Proprietary Data 180
    Safeguarding Government Classified Information 182

19. PERSONNEL SELECTION, MANAGEMENT, AND COMMUNICATIONS 185
    Personnel Selection 186
    Management and Leadership 188
    Communications 189

20. MARKETING COMMUNICATIONS 193
    The Perception 194
    The Elements of Marketing Communications 195
    Technical Papers and Presentations 196
    Marketing Communications and
       Business Planning 197
    Advertising and the High Technology
       Government Marketplace 199
    The Budget for Marketing Communications 202
    The Common Denominator 205

21. THE CHANGING MARKETPLACE 207

    Index 213

LIST OF FIGURES

Figure 2-1. Strategy implementation 18
Figure 2-2. Growth-forecast-risk relationships 21
Figure 3-1. Customers and marketing thrust 31
Figure 3-2. The total marketing system 32
Figure 4-1. Marketing manager organizational relationships 41
Figure 5-1. Technology base application to marketplace 55
Figure 6-1. Customer identifications 65
Figure 8-1. Market analysis relationships 84
Figure 9-1. Company and customer systems and the interface 92
Figure 9-2. Product life cycles: a closed loop 94
Figure 9-3. Expansion of core business 97
Figure 10-1. Key event schedule windows 114
Figure 14-1. Asset allocation to business opportunities 155

Figure 15–1. Interface for the new business acquisition plan   161

Figure 20–1. Progressive advertising campaign layout   201

Figure 20–2. Publication sleection and ad insertion plan   203

HIGH TECHNOLOGY MARKETING MANAGEMENT

CHAPTER ONE

HIGH TECHNOLOGY MARKETING MANAGEMENT

An Overview

A brochure describing a two-day seminar in Washington, D.C., had promised a complete overview of current technology, how that technology was being transformed into practical use, and how it was being marketed by the various commercial organizations represented by the speakers. This promise convinced me that I should attend because my own company was heavily involved in advancements in technology and the conversion of new technology into practical applications satisfying current stated needs. Furthermore, I was employed by my company as a technology marketing representative. I was disappointed. The speakers provided no additional information than what was already common knowledge in the marketplace. The promise of the brochure was not fulfilled.

In this overview, I summarize briefly what this book is all about. I concede that I don't have all of the answers. In fact, many outstanding marketing managers throughout the industry I address could argue successfully that I have missed important points. True, I may have overlooked some specific discipline they have followed for years which has led their company to a high level of success.

Every company has its unique organizational structural, and while what is shown on the published organizational chart has a ring of similarity across the total industry, there are many operational differences. Generally speaking, many marketing organizations operate on a day-to-day basis without full consideration of where they have been, where they are, and where they are going with their current game plan.

Short-term thinking is easiest; thus, marketing managers (and managers for other company disciplines as well) tend to be more concerned and knowledgeable about current activities. But high technology business requires intermediate- and long-range thinking as well. If an enterprise is to grow and remain profitable, strategic planning from one to five years into the future is essential. And, those marketing managers who carefully study technology trends and relate these to their own businesses will probably achieve their objectives earlier and with far better results.

The chapters that follow will delve into the complexities of this total package in order to describe a series of disciplines seen in most successful organizations.

The recommendations I offer are based on hundreds of evaluations ranging from those provided by company chief executives to the newly hired technician working in the laboratory, in industries large and small. From this mass of information provided by industry, and also by the customer, I have isolated many definitive guidelines you can implement without restructuring your marketing organization or the assignments of your personnel.

This, then, is the objective statement for this book. This is the promise—nothing more, nothing less.

THE DEFINITION OF HIGH TECHNOLOGY

What is high technology? In their overview for "Major Science and Technology Issues,"* the staff of the U.S. General Accounting Office stated that technology may be broadly defined as the knowledge, skills, methods, and techniques regularly used to accomplish specific practical tasks. They further state that prior to the twentieth century, technology was primarily the domain of inventors and craftsmen whose understanding of the scientific principles underlying their innovations and inventions was often "recipe knowledge"—they knew what worked but often had little understanding of why. Modern technology is often tied much more directly to scientific advances, although the lag between discovery and application remains quite long in many fields.

In the context of this book, I will address technology in terms of its practical application and the role of the marketing group in translating technology base into practical product applications with which you can address marketplace need. *High* technology consequently describes the segment of technology considered to be nearer to the leading edge or the state of the art of a particular field. It is that technology inherent in today's newest products. It is that technology emerging from the laboratory into practical application. However, it is also that technology held in the laboratory because management has yet to bring it out for its practical application although need may have already been expressed.

Many organizations are forever trying to define for themselves what technology is doing and how they can better apply new concepts to their own products. The objective is simply to get technology into a form that allows its effective addition to current products or the development of some new products to fulfill the need of the marketplace. There comes a time when pragmatists must be allowed to exercise their authority over research and development activities within an organization, otherwise the evolution process will be delayed (or totally halted) and your competitor will satisfy the need ahead of you.

MARKETING AND HIGH TECHNOLOGY

It would be cumbersome to spell out high technology and marketing relationships for every single field. In any study of tech-

---

* (A study by the staff of the U.S. General Accounting Office, "Major Science and Technology Issues," PAD-81-35, January 30, 1981.)

nology we are confronted with a broad and complex array of old, current, and new and emerging technologies. High technology has become a common term and is now associated with energy, the social sciences, the environment, genetics, medicine, national defense, space, and perhaps hundreds of subareas below these and other major headings. It is difficult to incorporate all of these disciplines into a single text to describe the technology and the management of an associated marketing effort. Volumes would be required. However, there are some general principles with respect to technology and marketing which can be discussed for the broadest applications. To technologists and their marketing counterparts not specifically addressed in this book, my apologies. Nevertheless, I feel that most of the principles discussed here have a sufficient across-the-board application that technologists and marketers in all but a few specialized areas will be able to visualize the relationship to their own activities.

I have selected the aerospace, electronics, and telecommunications fields as a baseline for these guidelines. I will use the government marketplace (including their prime contractors and all levels of subcontractors) as the basis for the discussion of marketing technique. It is this technology area and the associated marketing activity which is more complex, requiring a more planning-intensive approach. Many of the disciplines and techniques described here can be tailored to fit the particular size of an organization, other technologies, and other markets.

I believe the contents of this book will prove to be beneficial to almost all marketing managers, technical marketers, and others who hold similar job titles. Large organizations having the need for equally large and often grossly inefficient new business acquisition groups can and do assign many different titles to their marketing (sales) personnel. In contrast, the small business organization will often group organizational functions together and assign them to a single individual. Combinations such as engineering and marketing (sales), contract management and marketing, and operations and program management with new business acquisition responsibility are not all that uncommon.

However, regardless of the size of the business, the basic problem of organizing for and properly pursuing the new business available in the selected marketplace is more of a common problem than most observers are willing to admit. There are simply a handful of specific operational disciplines which, in my view, must be pursued if a company expects to achieve its growth projections. The size of the business enterprise is not a factor.

As we proceed, it is important to again keep in mind that although my vantage point is that of a long-time employee for a large business involved in government contract activity, much of the general guidance provided is directly applicable to all businesses involved in the high technology product and service sector. The scale changes in terms of dollar volume—or however else you care to set the scale—but almost all of the basic information remains applicable. The question simply becomes one of tailoring to fit the scope of your particular operation.

MARKETING THE GOVERNMENT CUSTOMER

The concentration on the government customer, his prime contractors, and the several levels of lower-tier contractors serving the government marketplace does not mean that high technology marketing is not practiced or is not essential between and among industrial organizations where the end product may be destined for nongovernment use. However, the needs of our federal government drive technology more than any other segment of our society. Funding for the progress of technology is derived largely from government needs—whether defense-related, nondefense related, or social.

Expenditures for defense are considered by many to be far and above what we need for a stable deterrent force. Although this may not be the proper forum to discuss these issues, defense spending creates a favorable ripple-effect throughout the economy. The Department of Defense estimates that about 28,000 jobs are created for every $1 billion they spend for material and services. They also estimate that an additional 7,000 jobs are created in the nondefense sectors of the economy as a result of spending by workers employed in defense-related business.

Military spending spawns entirely new and viable civilian industries. Examples are the fields of computers and software, and jet aircraft and communication systems employing geostationary satellites. Jacques S. Gansler, a former deputy assistant secretary of defense states in *Nation's Business*: "In each case development began with a perceived military need and defense research and development money brought them along until civilian markets were created."[*]

The excitement of high technology can be seen in nearly every sector of our economy. The forward movement of technology is so rapid that we are witnessing the age of specialists for every facet. No longer is it possible, in my view, for the general engineering graduate to function in an across-the-board design, development, or applications role.

A new era of integrated circuits is on the horizon which will bring about higher speed operation, greater capacity, smaller size, and less power consumption. The marriage of communications and computers is now a reality. Defense spending points to even more sophisticated weapon systems—many of them operating without human intervention once they have been programmed to perform their mission. Computer-aided design techniques have spread to include manufacturing, test, and quality assurance. Robotics has become as essential element in the development of the production lines for nearly all high technology products. These are indeed exciting times for the sales engineer, for the marketer, and for the marketing manager.

While it is true that large and small companies make major investments of their own in the advancement of technology, the need in the marketplace should govern how much money is invested and the tasks to be funded. Organizations making major investments in a strictly commercial venture very often have one eye focused on the government marketplace. I believe it is not necessarily true that many major and traditional government marketplace suppliers maintain a sustained focus on the commercial marketplace, although there are some outstanding examples of products available within nongovernment commerce which emerge from previously funded government programs. Some very successful companies serve both the government and the commercial marketplace because their products serve a bidirectional need.

[*]*Nation's Business* (May 1982):82.

MARKETING OR SALES—OR BOTH

The high technology company organization responsible for the acquisition of new business is usually identified under some marketing-related banner. Such departmental titles as business development, sales, marketing, and others which are less descriptive are employed. Regardless of the name, the function is defined as one of acquiring the new orders for a company's current products and the development of the company's new business involving similar products reflecting similar technology. In many companies, marketing is a function within a business development group. The implication is that marketing deals more with company representation in the marketplace while business development embraces the total spectrum of new business acquisition, including market analysis, competitor analysis, technology and technology base assessments, planning, and, in some instances, overall management of product lines.

There are nearly as many definitions for marketing as there are people working in the field. Marketing differs from sales in that the former deals with the total process of developing the market as well. This book is not written for the salesperson—except when it may serve to move one's career toward marketing and management.

I am not specifically dealing with what is commonly called the *consumer market*—not that high technology isn't applied to consumer products. There are probably thousands of books and pamphlets available describing marketing and sales techniques for consumer products. Some are of value to the high technology marketer and marketing manager. In this book, however, we address a much more complex and demanding marketplace. The differences will unfold as you read the chapters which follow.

In my opinion, the basic difference is the elapsed time between initial customer contact (the recognition of the sales potential) and the booking of the order. Perhaps this is where the real meaning of *business development* is brought into focus. It is a process during which the basic need expressed by the customer (and it is most often simply a statement and not a descriptive specification) is folded into a long-range company plan to fill the need—to capture repetitive sales once the need is satisfied by the new product.

One additional difference I should note here relates to compensation. Marketing (business development) personnel in nearly every high technology industry surveyed are compensated with basic salary only; no sales volume bonus or other incentive payment is provided. Compensation for senior marketing personnel (highly skilled) is set at high enough levels to attract and hold these employees without offering or awarding supplementary compensation. The reason is basic. A marketer may work toward a contract award milestone for a single program for one, two and perhaps even five years. Under such conditions of true business development, the marketer would soon lose interest if compensation were based on capturing the order.

Marketing high technology and products to government and industry requires a very broad comprehension of the total company approach to business. It requires a full understanding of the technology in terms of what has happened before, what is happening now, and where it is headed. It requires the best possible understanding of the customer, the competitive nature of the marketplace, and one's competitors. Marketing also requires the ability to read trends and indicators regarding all of these segments. And, marketing requires the ability to adjust a game plan when the trends and indicators tell you to do so.

For high technology to fill a stated need expressed by a government agency (Department of Defense, Department of Transportation, NASA, and others), the gestation period may be measured in years. Two reasons are most obvious: (1) the application of current technology to some complex need requires a methodical phased process; and (2) the need may seek an operational performance which cannot be filled by current, state-of-the-art technology.

Technical marketing, then, is the process of fitting together the technology of the business enterprise and the expressed need of the high technology marketplace. It is a cumbersome process requiring broad technical, sales, and management skills.

THE ACTIVE ROLE IN MARKETING: THE SYSTEMS APPROACH

A study of marketing techniques in high technology industries over the past several years has convinced me that success (realizing most of one's objectives) is best achieved by the marketing manager approaching his task from a systems viewpoint. It is my feeling that marketing managers often perform specific segments of their assignment efficiently, but many lack the ability to tie their total activity into a system (a marketing system) which includes the management of the interfaces. In most high technology industries the complexities of their companies, the marketplace, and their competitors are so great that unless some organized management effort is undertaken, job market opportunities may be missed and thus company objectives may be unfulfilled.

Within the high technology aerospace industry and others, I see a common thread weaving through successful marketing organizations. It is the management of the company technology base. Even the best marketing organization will fail when the technology base crumbles. Unless their engineering groups are translating high technology into products satisfying marketplace needs, there appears to be little a marketing group can do to provide for the growth every company wants to achieve. Yet, marketing shares the responsibility for that failure.

I will review many techniques for marketing managers planning for the future. Many don't analyze, but instead only react to what has already happened. They don't strive to form their segment of the market, but instead participate in what others have already defined. Also, with respect to planning, managers may become so embroiled in today's issues that planning is relegated to a staff function. The cancer of mediocrity soon attaches itself to the body of the organization. Consistency and concentration— these are essential attributes for any organization.

You may win a coveted new contract because of who you know and not by virtue of submitting the best cost and technical proposal. You may even win because you compromised rules of conduct and twisted the arms of the procurement group. Eventually, these successes will no longer be yours and you will be forever struggling with the often-heard question in marketing groups: "What went wrong?" I believe that answer is that you are not managing the marketplace. You are not managing the company–customer–competitor business environment.

Some of my conclusions may be regarded as somewhat irreverent by my colleagues. They may say that I grew up in that same community, that I had the opportunity to witness the good and the bad, and now I treat the subject in a rather critical manner. However, having been in business for so many years tends to give the careful observer (the listener) the opportunity to identify the best trails to success, and to begin to understand that when you obey the key rules of past success you can help you company show growth and profitable operations year after year.

Success is measured in many different ways. You can grow with your company. If your company is successful, you can be successful. But, company success is, in my view, the result of the contribution of its employees. And, that includes the marketing managers.

In evaluating marketing functions, one will often encounter elements of an evaluation criterion covering the active role in the marketplace versus the passive role. Managers may disagree on the precise definition of these terms. An active role is really the pursuit of the new business acquisition process by preconditioning the marketplace so that when actual procurement of the product reaches contract stage, you have already performed almost all of the tasks required to win. The passive role is simply standing aside and waiting for the elusive call to perform. Almost every marketing manager in the high technology business environment operates in the active mode. Some are much more aggressive than others. Aggressiveness is a path to success. Managed aggressiveness in the marketplace is essential to long-term growth for your company. The term *managed* implies control of the methodology for addressing the marketplace. That is what this book is all about.

CHAPTER TWO

COMPANY STRATEGY AND PLANNING

The marketing manager in a high technology industry serving a government customer is facing a most difficult task when confronted with the requirement of preparing his portions of a strategic business plan. Due to the variables associated with current business, technology base, customer and product mix, and general business conditions no two companies address the issues of strategic business planning in exactly the same manner. However, throughout the high technology industry, there appear to be many basic considerations which cannot be overlooked in business planning. Basic issues such as company mission and objective need to be addressed periodically if only to confirm the posture of the company. A mission statement is only a stake in the ground, but it may establish the direction of the company and the framework for all near- and long-term business planning.

This chapter will review many of the basic considerations for the company business plan and the role of the marketing executive. Business planning cannot be an annual affair. Every single business decision throughout a fiscal year must either be a reflection of the business plan currently in force or an approved modification thereof. And, the decision processes and recommendations out of the marketing segment should be continuously tested against the plan for validity and continuity of the business mission.

DEFINING THE TERMS

When discussing company mission, objective, strategy, planning, or any similar term, it is imperative to achieve a common understanding regarding the meaning of the term being used. What is a strategic mission planning objective? It could mean nothing at all. Many terms often become standard within a company management hierarchy. You see them repeated over and over again. For our purposes, we define mission as the purpose of the business and the products and services provided. Objective is a condition of the business which management wishes to achieve at some future point. Strategy is the general plan to meet an objective or set of objectives. Planning is the development of the detailed, formulated scheme for accomplishment. There are dozens of variations of these terms and definitions. Often, companies will establish their own precise set of terms and definitions so that everyone involved in the management of the business and planning for the future will have a common baseline from which to operate.

Most business planners will tell you that one of the first steps in the successful business planning process is the definition of what the business *is*, what the business *will* be, and what the business *should* be. Some of what the business is today should probably be discarded, since analysis may show that it is sapping the strength of your business in other areas. In other words, if it no longer provides positive cash flow, discard it. Out of this kind of analysis will emerge the early sets of objectives for the business. The next step, then, is setting the strategies to meet these objectives. Businesses can have a single objective or multiple objectives, a single strategy or multiple strategies.

Business planning is the periodic restatement of objectives and strategies and the establishment of programs to implement the strategies. Simple statements of business objectives are of little long-term value by themselves. They look good on paper, but their value in terms of contributing to company profitability and growth decrease to zero in a very short period of time. Strategies and their implementing plans lead to achievement of whatever objectives you set.

I have devoted much of my career to understanding what it is that companies strive for. Why are they in business in the first place? What is it that separates mediocrity from success? What is it that compels company management to work so diligently to achieve whatever objectives they have set for themselves? What are the key objectives of a company that will set it apart from so many others that show only mediocre results? The two key objectives, I believe, are *growth* and *profitability*. All of the other objectives one can list—and they are many—are (or seem to be) secondary to these two primary objectives. In fact, I have argued this point with many of my colleagues. The point is that unless these two objectives are at the very top of the list, then the business enterprise may fail within a few short years. Embodied in these two objectives are all of the other objectives a company strives to achieve. Business decisions which do not contribute to these two objectives are probably not good decisions.

THE BUSINESS PLAN

In examining multiyear business plans for over two decades, I have yet to find one which did not project growth for the company involved. In many instances this is nothing more than a short-term solution to a long-term problem. Companies just don't project a downward trend in their business for very long. A manager not projecting growth for his company will usually be replaced by one who will. Realistically, managers cannot expect current business activities to fill the total volume for the growth projected. Growth comes from the business planning activity. In high technology business, today's products become obsolete (lose their appeal in the marketplace or no longer fill the need) in a very short period of time.

Successful managers (and companies) maintain a combination of business activities identified simply as *diminishing, sustaining,* and *emerging*. In many companies you will find plans dealing with the divestiture, or phase-out, of products and product lines. The discontinuance of marginally profitable activities—where market size is diminishing—is good business practice. Discarding old products is a difficult process, but it is essential for good business planning. I have seen examples of products that long ago lost their appeal in the marketplace being retained by companies because they couldn't face discarding them until the very last sale. Often, they have replacement products of their own that form a part of the sustaining product category. The dissolving process will free assets which can be used to further develop the sustaining business activities and nurture the new products ready for release from the laboratory.

The purpose of business planning is to force a study of the total business environment. Most successful companies maintain a continuous business plan process and monitor progress against the published plan, also on a continuous basis. The high technology business environment is constantly shifting and changing. Whatever your assessment may be today, you need to be on a constant alert for change.

The absence of planning for the near as well as long term will allow your company to drift along with the trends in the marketplace. A company without a business plan has no rudder and will eventually drift into obscurity. The questions, of course, are how much business planning should be undertaken, how often should a plan be updated, and how detailed should it be? And, what is the role of the marketing manager in the business planning process?

Give me a straight edge, a pencil, and a sheet or two of graph paper, and I'll give you a five- or ten-year plan for growth that will look impressive in any long-range planning document. I'll tie it back to spending projections for your products, show a percentage for investment, compute and lay in return on sales, return on assets, and other indicators, and (on the surface) convince many chief executives that I've really got things under control. If it's done in color with some sort of a three-dimensional effect, it'll sell even better. But, what does it say about the real company? Probably very little.

Even within my own company, a monstrous paper transmittal flows from our corporate offices and senior level executives each year to describe detailed requirements for preparation of the annual five-year business plan. Hundreds of managers throughout the world of ITT react (some in terror) to the requirements. I've often thought that a simple statement from the chief executive officer, such as "prepare me a detailed plan to describe how you will conduct your business for the next five-year period," would be a refreshing change. However, hundreds of plans, each in a different format, are difficult to assess. Instead, if they are arranged according to directives, they can be placed side by side and added across to provide the total company business outlook and plan. While it may be as complex as any found in industry, the ITT planning system is effective and yields a thorough annual examination of the total business.

There are valid arguments for and against formal planning groups within a company. Smaller companies cannot afford a separate planning group, and thus must rely on diverting personnel from other projects and programs whenever the business planning function is to be staffed. Smaller companies, and in fact many larger organizations as well, utilize one or more of the commercial market analysis services to do their general and specific research for them.

Personnel diverted from current programs to perform some business planning function are usually so involved with solving their own project problems that they pay little attention to the planning process other than to fill in blank spaces. This is not good business planning. Engineers and marketers cannot be expected to suddenly shift their emphasis from on-going activities to business planning. Most often it is the on-going effort that receives their attention. Business planning is only a temporary delay in accomplishing their primary assignment. What is really called for is the dedicated planning function staffed by people who are not distracted by the problems of current activity. Their only function is to perform the required research, data analyses, and the summarization necessary to equip senior management for the task of developing the company objectives and the strategies to fulfill those objectives.

Injudicious planning disciplines are routinely practiced by many companies because they have not examined prior plans in light of the current posture of the company. Some of the most revealing shortfall in business planning is isolated by comparing actuals with plan. The objective is to determine how well the planning function is operating. Pitfalls are scattered throughout this process. In looking at actuals, one has to be able to identify departures from plan and accurately identify the basic reasons for the departure. It is often extremely difficult to pinpoint shortfall in planning, particularly when the assessors either formulated the plan or are responsible for the conduct of the programs that would have fulfilled the plan requirements. They tend to be less than objective. However, this internal grading process must be conducted to raise the level of the planning process.

To many, a long-range plan is measured in months—not years. Planning for the long term can get lost in the noise of the marketplace and the absolute requirement for achieving current year and current year-plus-one budgeted performance. It may be wise to split the functions of a business planning group. One group may be dedicated to achieving the current budget (plus one year) while a second is dedicated to the long-term plan. The interaction between the two groups must, of course, be constant. The long-range planning group, unless tethered to current activities (successes and failures), may not visualize the continuity of the business and embark on a path that does not have the essential relationship to what the company has achieved and can profitably achieve in the future.

A current year business plan containing out-year projections may be only a reflection of a single reading of your business environment data base at one point in time. If you publish a multiyear plan on an annual basis and your input to the plan is based on your business environment assessment completed just prior to its publication, the plan has minimal value. Business planning should be a continuing process. If you are constantly aware of developing trends, current issues, current and developing needs in the marketplace, shifting posture for your competitor cluster, and the condition of the company's technology base, then you are generally well prepared to produce a useful plan for the future.

STRATEGY IMPLEMENTATION

Mission statements and sets of objectives and strategies are of little value unless they are supported by a series of projects and programs that will implement them. The best possible company strategies will remain just that until management defines their programs and provides funding to support the implementing tasks. One can look at this portion of the business as a subsystem within the total business (see Figure 2-1).

Strategies must be translated into work packages. As Drucker* has said, plans remain plans unless they degenerate into work. Assets required to perform the tasks described in the work packages must be made available to personnel assigned to the tasks. And, a simple and effective control system should be established to monitor progress, assess directivity of the effort (is it on track), control the depletion of the assigned assets, and continually reassess to confirm that the original intent is still valid. Competitors and customers alike are constantly shifting their own emphasis, and since you are part of the overall system you must also act and react according to their actions and reactions. It is a closed-loop system.

PLANNING GUIDELINES

In many organizations the planning function amounts to nothing more than a pass across the crystal ball. Successful organizations and successful marketing managers believe the business planning process must start with the best possible data they can accumulate. It is often a reflection of how your management perceives their business and what they want it to become. The size of the data base from which they will make their decisions somewhat dictates what the decision is going to be. It has been

*Peter F. Drucker, *Management Tasks, Responsibilities, Practices* (New York: Harper & Row, 1974).

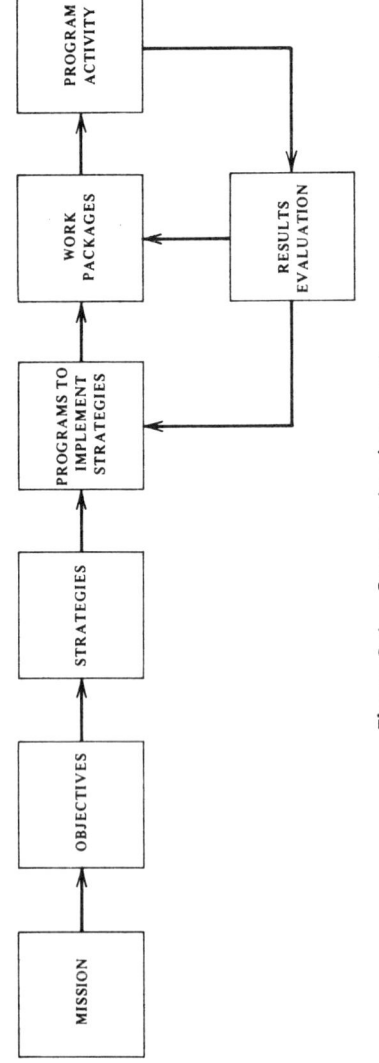

Figure 2-1. Strategy implementation

said that the information for a correct decision for a business is nearly always buried somewhere down inside the data base. If this is true, then the broader the data base, the better the decision that will come as a result of its study.

However, you should build in some escape routes to prevent you from becoming trapped in a plan that may lead you to failure. This course correction allowance—or the permission to change that is inherent in all plans—is essential in any high technology business. It is nothing more than the protection built into the plan to avoid the impact of surprise.

Any long-range plan must have flexibility built into it to allow adjustments in the day-to-day conduct of the business. It would be incomprehensible from a purely operational standpoint for any high technology business to become so strapped to its business plan that fine-tuning is ruled out. The nature of the marketplace demands flexibility. Let me cite two current examples: (1) the MX missile system, where development has been delayed because we couldn't obtain a consensus regarding basing and launch platforms; and (2) the B1 bomber, where the argument continues that we can rely on the B-52 until some new bomber configuration evolves utilizing advanced stealth technology. Here we see hundreds of millions of dollars involved in indecisions. Hundreds of prime, sub, and lower tier contractors and suppliers are left hanging on a limb, and thousands of employees who thought they might have employment security over a period of years are either looking for work or are wasting their talents on less complex projects. These may be extreme examples for some. For others, the companies involved in these programs may represent the difference between success and failure in terms of their financial condition. It is easy to imagine the gut-grinding processes they are undergoing with respect to their business planning. This segment of their marketplace is in a state of chaos, and there isn't much they can do about it except monitor the progress of the government's decision process on a day-to-day basis.

Long-range plans can be quickly evaluated for soundness by comparing current year results with the original plan for the current year. Wide excursions from plan usually indicate poor planning or mismanagement of current year activities. Managers who consistently complete a year's activity to within a few percentage points of plan are usually less suspect than the manager who shows either wide positive or negative swings. However, his conservatism and his ability to hit his target may also suggest a poor planning system. His success needs to be examined just as thoroughly as his failure. Major excursions from plan must be examined to determine cause. The government marketplace, as unpredictable as it is, is usually blamed for wide excursions from plan. But the astute manager will search his planning system to determine if certain trends and indicators were available at the time the plan was formulated, and, having read these signs correctly, could he have avoided the surprise of large negative and positive swings.

There are conflicting thoughts with respect to technological breakthrough and how you should save a place for it in the business planning process. Technologists serving almost all scientific disciplines strive to achieve breakthroughs whenever they walk into their laboratories. But the business impact of a breakthrough is not sudden; the marketplace does not adjust nor can it be adjusted rapidly. Technological breakthrough which would render current products obsolete requires a finite time to reach fruition. Businesses are not willing to scrap whatever products they now have just to address the untried and unproven. Business operations are geared to whatever they are now doing. Change is incorporated over a period of time. Besides, a new breakthrough coming out of the laboratory today is usually not ready for the practical application in the field tomorrow. The pragmatists need to devote sufficient time to the new technology to bring it to practical cost-effective application.

Business planners will often segregate technological status and conditions over increments of time and attempt to predict when the new will reach practical application. The transition out of the laboratory is often slow; the pragmatists also need lead time.

FORECASTING

Business planning usually concludes with a forecast for the future. Techniques that should improve your forecasting accuracy

FORECASTING

are covered in Chapter 10. But at this point, I want to show the relationship between growth projection and forecast visibility (see Figure 2–2). A company five-year plan projection from plan year 1 shows growth from an expanding business base. But, your

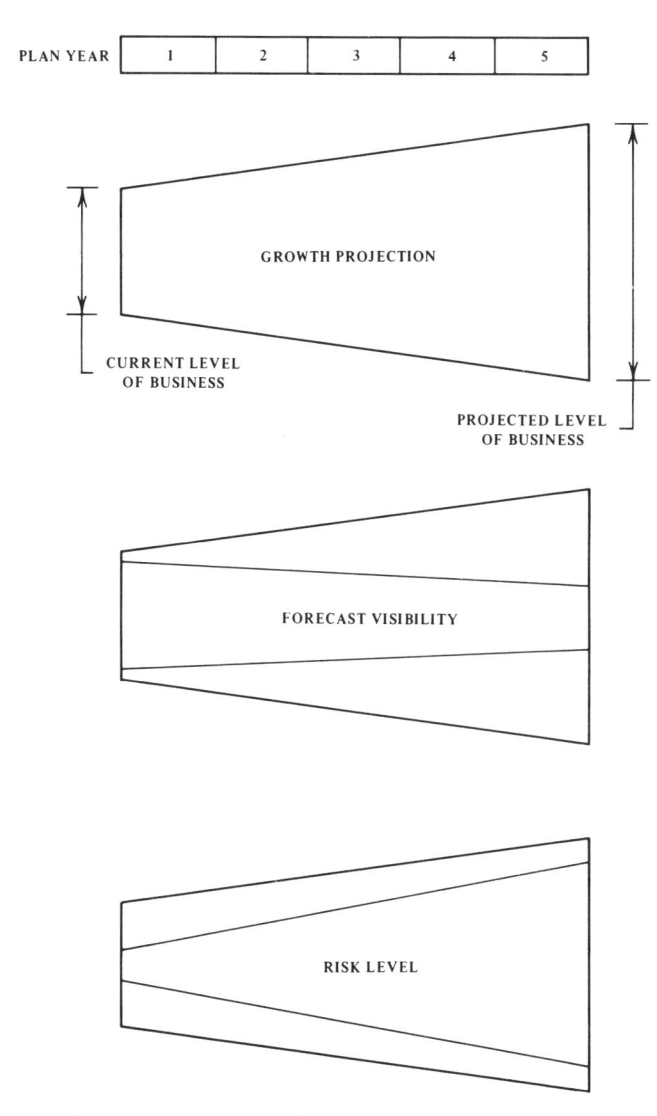

Figure 2-2. Growth–forecast–risk relationships

forecast visibility obviously decreases rapidly and may reach zero even earlier than the diagram portrays. The intent is simply to show graphically the reason for maintaining a continuing business planning process. The diagram should serve as a reminder that the forecast visibility wedge is driven right into the heart of the growth plan. Plan validity diminishes rapidly.

One additional element is added to the diagram in Figure 2-2—Risk. Your at-risk conditions expand rapidly over a plan period. The dynamic conditions of the high technology marketplace bring risk into play earlier and often in a much broader sense. While I won't dwell on risk assessment here, it is vital that the risk factor play a major role in good business planning. Good business planning reduces risk and is an element of risk management. Risk management includes good business planning. I have heard many managers talk about failure management, which is also seen as the ability to manage under the conditions of some major departure from plan. Where you have risk you also have the possibility of failure.

INVESTMENT STRATEGY

Sooner or later, you will get to the point in business planning where decisions will have to be made regarding the strategy for investment. In any successful business, some percentage of the profits needs to be allocated toward nurturing the business—a plowing back into the business of the resources required to fulfill certain requirements of the strategic plan. It is vital to reduce the speculative nature of the action to its lowest common denominator. No company has unlimited resources (at least I haven't found one); therefore, a selection process is required to pick the best possible elements upon which to heap whatever resources are available.

Throughout this text I will refer to technology base management. It is the process of maintaining the essential base from which a company can pursue its new business target opportunities for the near and long term. Each of the targets you have selected has a set of essential technological requirements. By comparing the needs of these programs to the attributes of your technology base, you can visualize the base shortfall and begin your planning for corrective action. In Chapter 3, I will plot this segment of your business to show the relationship to your total business system.

Investments in the technology base for your company are essential. There are many sources for the funds needed to accomplish base expansion; I will review these in greater detail in later chapters. The plan for investment of company-generated funds requires (and receives) greater attention than all other investment fund sources. Usually, when it is time to consider the needs of the technology base, managers should exhaust all other sources before submitting their funding requirements that draw away from a company's net income.

While there are many yardsticks for measuring the need and effectiveness of strategic investment (the investment now for the future health of the business enterprise), the last and most important hurdle is the expected return on that investment. This element should be the primary concern of the marketing manager and his staff. They have access to the marketplace; they must interpret new business potential. They can measure technology base posture and the need expressed in the marketplace. When these relationships are clearly defined and supported by the best possible data base, approvals for strategic investment are more readily accomplished. The role of the marketing manager is vital.

THE ENVIRONMENT

The social–economic–political environment plays a much larger role in business planning than many believe. Company management cannot ignore the implications nor the continuing impact of this area in the conduct of business planning. The control and demands placed on business enterprises by all levels of government have been discussed widely over recent years. What is viewed by many as a stranglehold on the free enterprise system may be lessening a little, but the *bigness* and *badness* of elements of our government, at least as interpreted by some business executives, are still a deterrent to the conduct of a successful business. Some outspoken critics believe the free enterprise system may be in jeopardy. I doubt that. Instead, we see experimentation with the system, but usually only to make it a better system.

Currently, with the trend toward less governmental support of various social programs for the disadvantaged and other segments of the general population, hundreds of organizations representing these groups are increasing pressure on the private sector to become more and more responsive to their needs. How all this relates to business planning is difficult to relate in a general statement which embraces all companies. The social responsibility of the business enterprise goes far beyond providing goods and services, a workplace for employees, and the annual payroll. A company's business mission should include a forthright acknowledgment of social responsibility. And, a business plan should clearly identify how that responsibility will be fulfilled.

It is becoming more and more evident that companies are expanding their planning functions to include a full assessment of the changes seen in every sector of government. The impact is not as direct as the cancellation or modification of a major procurement action; it is usually slow and difficult to define or quantify. However, to say it doesn't exist is a detriment to good business planning.

THE MARKETING MANAGER'S ROLE IN BUSINESS PLANNING

The marketing manager must play a leading role in business planning. It is usually his function to provide market data and the analysis of current products and technology base with respect to the need in the marketplace. *Need* in the marketplace is the key to the entire process. Defining the business strategy is simply the apex—the cap for all of the front-end work you need to accomplish in order to establish the basic framework for your business; or, what your business should be to accomplish the objectives you have set. Strategy doesn't identify *how*; it simply identifies *what*.

As we proceed through the chapters which follow, you will begin to visualize how management and control of your marketing function plays such a major role in business planning. In fact, it is recognized by most experts in business planning that marketing input is the key portion of the data base. Maybe that's why many major corporations refer to the generalized marketing function as business development. The title is certainly descriptive of the real function marketing performs.

A GUIDE FOR BUSINESS PLAN FORMULATION

Despite all the discussions up to this point regarding business plans and planning, there may still be some confusion regarding how plans should be formulated and how a marketing manager (and other managers as well) can improve their procedure for plan development.

The initial task is to understand the elements of a good business plan. Many companies have excellent outlines to follow; others just let their plans begin from what the company is today and then gradually expand on that base to what the company should be at some point in the future. For many, the roadmaps for moving the company forward from where it is today to some point in the future are ill-conceived and quite difficult to follow.

A business plan, if it is to serve a useful purpose, provides the essential—but basic—set of guidelines for the conduct of the business. If prepared with that objective in mind, the plan is a better plan because the preparation criteria have become rigid within the company organization. If a completed and approved business plan is not employed to guide the business—and this is a known and accepted condition within the organization—then the studies and the work effort associated with plan preparation will be superficial. You have to believe in the business plan function if it is to produce meaningful and useful results.

In developing your business plan, you must clearly define the current condition of the company in terms of technology base, finances, and position in the marketplace. This, in my view, is the first vital step. Then, you should envision your company two, three, five, or even 10–20 years into the future. What is the position you want for you company? Given today's posture and the inertia inherent in almost all organizations, what future objectives are realistic and achievable? In setting long-term objectives, I find many aggressive company management organizations utilizing most of these criteria:

general market trends and indicators
market share increase potential
new market penetration potential
existing product evaluations for improvement or for discard
investment initiatives
capital requirements
personnel requirements
facilities and locations
productivity enhancement
divestitures
acquisition plans and programs

Now, with these two positions established, the remaining task (and the most vital) is development of the plan to move the company from today to the position reflected in the objectives. It is this *interface module* that is the heart of the business plan. Companies produce excellent evaluations of their present posture and exciting statements concerning posture in the future. The real measurement, however, comes in how well the *plan for movement* from point A to point B has been formulated and whether or not all (or most) of its elements can be implemented. This interface module eventually yields work packages which are developed for the single purpose of moving the company ahead toward its objectives. (Refer again to Figure 2-1.) This is the most basic and essential element of good business planning.

CHAPTER THREE

VIEWING THE MARKETING FUNCTION AS A TOTAL SYSTEM

In any complex system, performance of the whole is achieved through the required functioning of the many interrelated subsystems. Such is the case with high technology marketing when we address the function as a system.

To gain the fullest understanding of the complexity of the high technology marketing function, you should consider the development of a systems approach wherein each of the subsystems within your company, and within your customer and competitor environment, are interconnected in the traditional systems block diagram format.

Treating high technology marketing as a system may be a new concept for you. However, marketing managers are expected to develop the right recommendations under complex and varying sets of conditions. The systems approach may provide the clarifying set of instructions. At the very least, the systems diagram will provide a checklist for testing the decision process related to the marketing activity. The diagram will provide a framework for the total company approach to the conduct of your business.

High technology organizations and their personnel are accustomed to using such diagrams and techniques in their laboratories. The emergence of systems analysis techniques from the technical environment into management sectors of business is not new. A systems approach for the marketing management function is an essential first step in moving away from some of the traditional management concepts and toward a more effective understanding of the complex and dynamic high technology marketing environment.

THE SYSTEMS OBJECTIVE

The objective than is to accumulate the essential characteristics of your business and your business practices and to portray their interrelationships in such a manner so that you can easily visualize the total system in which you operate. Thereafter, when evaluating a continuing, modified, or totally new business thrust, you can scan your total system to determine the impact of your thrust, locate the hard and soft spots within the system which need to be addressed, and establish—in your own mind—the likelihood of success or failure under various conditions.

There are many advocates of the technique of looking at a business from a total systems viewpoint. It is similar to the methods employed by high technology project managers for examination and study of a total program in the laboratory. In the context of looking at your company operations as a system, we include the total company operations and the total business environment in which the company operates. Most marketing managers are exceptionally well versed in the external environment of their business. They are totally immersed in the conditions of the marketplace. However, many of them have less than a full grasp of their competitors' status or the conditions within their own department and company. When these separate environments are brought together under a systems umbrella, the new business acquisition process is greatly improved.

The systems approach to high technology marketing is intended to obtain an essential balance across all facets of the business. Emphasis is then applied where it is required. And, as the business moves forward, the needs of the systems elements will shift in emphasis. Each new business thrust brings new challenges to the engineering, proposal, and manufacturing groups. Armed with a total systems diagram, the marketing manager can better assess risk areas, place emphasis where it is needed, and thus develop a total company posture directed toward the new business acqustion process. Before we address the development of the systems diagram for a typical company, an examination of certain systems analysis concepts may be helpful.

We must first identify the boundaries of our system. Here, our boundaries are set by the company, the competition, and the customer. Throughout this discussion, the terms *customer* and *marketplace* are considered interchangeable. This somewhat simplifies the development of the diagram. Obviously, the conditions surrounding some specific customer may or may not reflect the true conditions of the total marketplace, but this inaccuracy is not detrimental to our understanding of the concept. You can adjust your own diagram to reflect your understanding of the marketplace addressed by your business.

It is also essential that we establish the purpose of the system. As stated previously, our purpose is to better visualize the elements of the total business environment, but from the viewpoint of the marketing manager.

In systems work, levels of abstraction are usually established beforehand to aid the analyst, but in our example (and the purpose for which it is constructed) we should portray and thereafter regard all elements of the system as concrete. A higher order discipline may include model construction and simulation, but such an operational concept is not commonly employed within the marketing function in the industry we address here. The variables within the high technology marketing environment are complex. Translating all of these variables into a language suitable for modeling and simulation is much beyond the scope of this text.

In any complex system the individual performing elements are made interactive within the system through an often equally complex interface subsystem. Thus, if one concludes that marketing management is the interaction of many individual performing elements, then we can conclude that marketing management must also construct and manage the interfaces in order to be effective.

Throughout this chapter and those which follow, we will draw your attention to the systems aspects (the functioning elements and their interfaces) of marketing management. Unless each systems element contributes to the mission of the system, then it is either not essential to the mission or its function is so poorly defined and controlled that mission success may be jeopardized.

SYSTEMS DIAGRAM CONSTRUCTION AND USE

To begin our examination of the systems concept in greater detail, we have selected a medium-size company dealing with high technology products in a somewhat complex marketplace. The company's market is a combination of direct government agency sales as a prime contractor, and a substantial mixture of sales to other prime contractors selling to the government and to subcontractors who in turn sell to prime contractors. By definition, then, our representative company covers the total spectrum of sales throughout the high technology market they address (see Figure 3-1). All of these customers may be regarded as either military or commercial.

Figure 3-2 depicts the systems diagram for a representative company. You will notice we have intentionally called your attention to the block entitled "Customer Analysis." The reason for this is—or should be—obvious. Most of the analysis work performed by the high technology marketing manager relates to the evaluation of business opportunities.

There is no intent to show the detail of each of the blocks

SYSTEMS DIAGRAM CONSTRUCTION AND USE

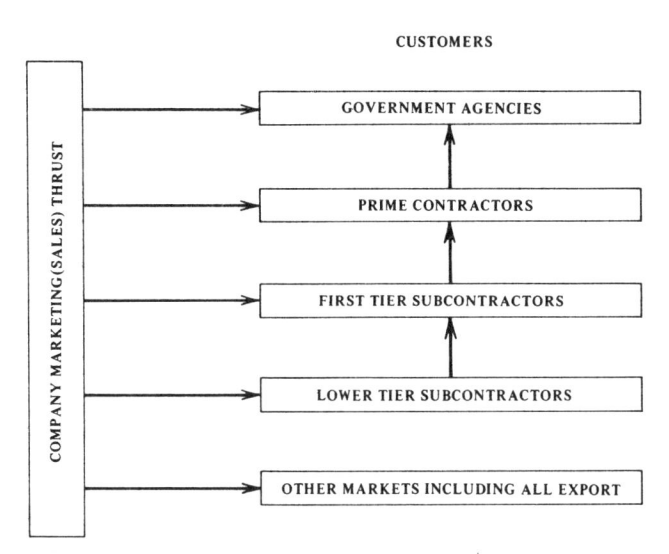

Figure 3-1. Customers and marketing thrust

forming the diagram. The purpose is simply to identify all the elements of the system and show how they are interconnected. The detailed descriptions will be covered in subsequent chapters. Our purpose here is to identify the total business environment for testing the decision process related to new business opportunities. Let's examine a few examples so that you may reach a better understanding of the concept.

The block entitled "Elements Contributing to Technology Base Formation" should include a listing of programs or products. Further descriptions are not necessary. The listing makes them part of the total systems diagram.

The basic and essential company doctrine is depicted in the segment of the diagram identified as "The Company." Every company should have sets of rules to serve as primary guidelines for current activity and all short- and long-term planning. These guidelines are usually reexamined on an annual basis to correct the course for total company operations. They become visible through the business planning function, provide company management with confidence in planned operations, and guide line and staff managers in the day-to-day operation of the company.

Figure 3-2. The total marketing system

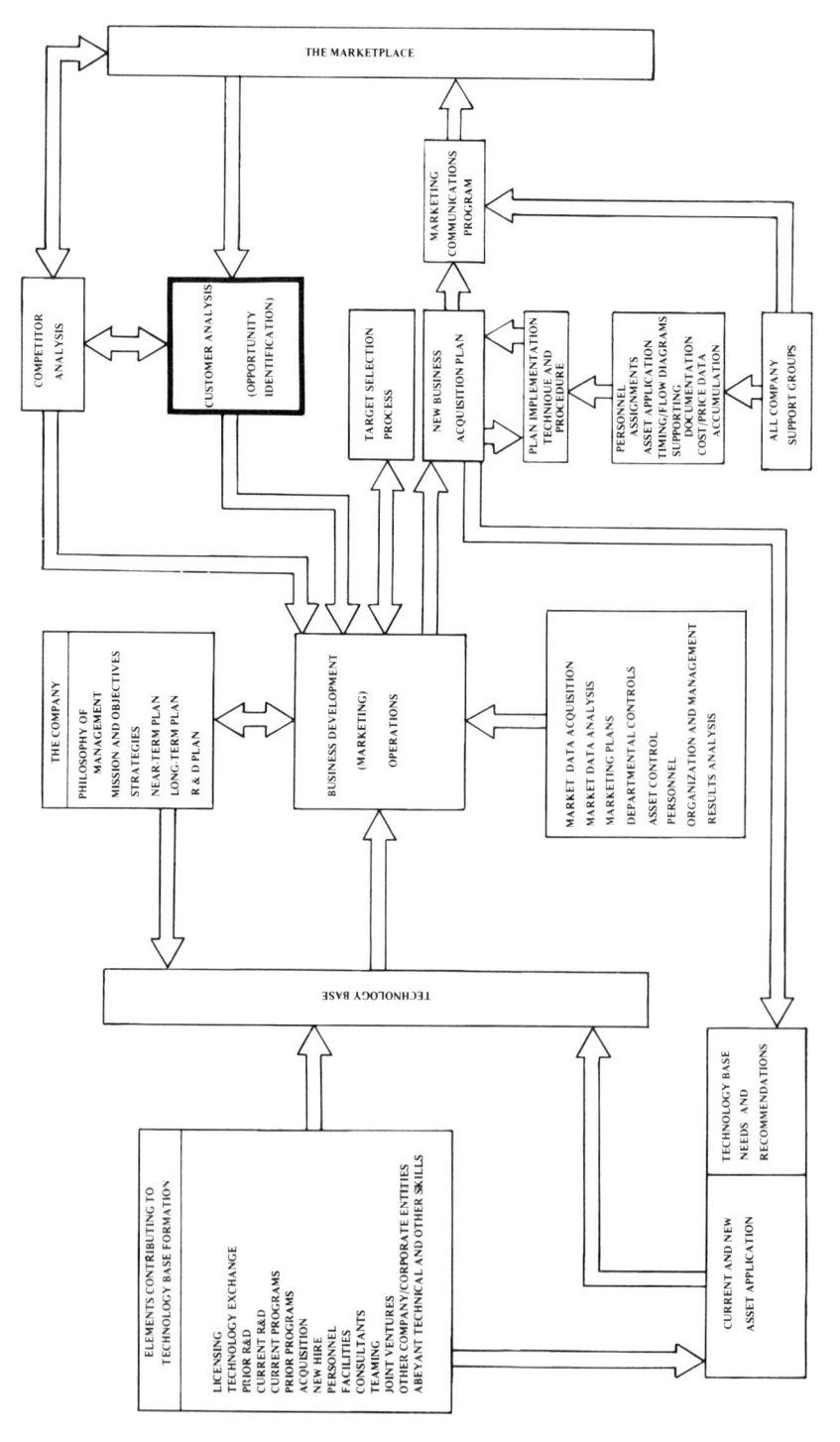

The value of the systems diagram becomes clear when you use it for the first time to evaluate a new or different thrust in the marketplace. As an example, let us assume that a government agency has determined that a need exists for new equipment. For our purposes, we'll refer to the new product as a widget. Widgets usually aren't very complex products, but let's assume that our widget is an electromechanical array that performs a missile control function for a new fighter aircraft now in development at the ABC Aircraft Corporation. Our widget would be procured by the government and furnished to ABC for installation in the aircraft. The company winning the contract for the widget would therefore be a prime contractor to the government, but would also be required to work closely with ABC to insure that aircraft interfaces are properly considered during the development phase.

The company marketing group became aware of this developing need through a routine contact with the agency. The newly assigned government project manager—not yet constrained by procurement rules which would prohibit him from discussing the project freely with industry representatives—laid out his program plan, showed a projected funding plan, and described how the widget would be employed. He also provided preliminary specifications for the device. During the discussions, the marketing representative learned that a 12-month development phase would be funded to determine feasibility. This would be followed by an engineering development program of 18 months, a combined contractor–customer test and evaluation program lasting four months, and then production of between 500–800 sets. The total quantity to be produced would depend upon the number of aircraft funded. The preliminary unit cost for the widget had been set at $10,000. The early estimates for the development phase did not exceed one million dollars.

In examining the preliminary specification, the marketing group determined that the requirements for the widget seemed to fit the capabilities of the company because of production of similar types of equipment. Furthermore, one segment of their current research and development program was expected to yield a sufficient new technology base so that competitive technical and cost proposals could probably be prepared and submitted.

With the preceding brief description of the proposed project, what should a marketing manager do to determine whether or not the project should be added to his list of proposed new program acquisitions?

Before we proceed with the description of how the systems diagram comes into play here, let me say that the widget program information just described will usually not be sufficient to enable you to make any decisions. One meeting with a customer contact is simply not enough research into a new program. In Chapter 9, we will fully develop this subject of target selection. The purpose here is to start you thinking in terms of applying new-found business opportunities to the diagram to determine applicability to your current business plan, and whether or not you can compete in that specific market for that specific product. Also, try and determine whether or not it is the kind of business you need to acquire to fulfill the long-range aspects of your business plan. Just winning the order for the widget may be insignificant if it does not lead you toward the company's objectives. The new contract must also fit into and become a part of the total plan for company growth and profitability.

Remember, the purpose of the systems diagram is to give you a roadmap to follow in your decision process. As you proceed through this process for the widget, you can stop at each of the subsystems in your diagram to evaluate how the program fits into your overall business plan, how well you are prepared to add this new program to your list of new business acquisitions, what new technology base requirements you will need to achieve the competitive level in your technical response, which competitor organizations you need to be concerned about, and whether or not your company will need to make an initial investment of their own research and development funds in order to achieve the required technology plateau. And, how does the widget program contribute to the mix of business opportunities you already have on your listing? Does it fit? Does it also contribute? Is it a program having technical content which is essential to some other program you now have on your new business opportunity listing? If you lose the widget program to a competitor, does that endanger some other program you are both pursuing? Will the competitor now have the advantage? Must you win the widget contract to keep the competitor from intersecting your business plans?

These are questions that only you can answer. I don't have the answers for you. No two company organizations or their mission and objectives are exactly alike. No two new business opportunities are exactly alike. No two company management structures are exactly alike. You must evaluate the widget opportunity depending on how you perceive your situation. My purpose here is to show you how the diagram serves as an invaluable aid in the decision process. As we proceed through the chapters which follow, we will periodically focus our attention on the diagram again to show how the specific subject we address fits into the overall diagram and the systems concept we have discussed here.

VALUE OF SYSTEMS APPROACH

Now that we've examined a new program opportunity against the diagram, it is probably a good time to evaluate your own decision process regarding new program assessments. Perhaps you already have the basic framework for your own systems diagram. You'll never appreciate the value of this tool unless you develop your diagram for your operation and then test it for its usefulness.

The diagram also serves other purposes. Assessment of a new business opportunity is only one. Marketing managers usually have a strong voice in the development of the company research and development program. In our diagram, note the relationship I have drawn between company technology base and other elements in the system.

Like so many other marketing specialists, I have attended dozens of seminars and explored perhaps as many texts dealing with marketing and management principles and their application. However, the bottom line always seems to be the inability of the pragmatic manager to implement the principles and disciplines reviewed by lecturers and writers because of the inertia of his functioning organization. He may at best modify only a fragment of his total operation because to do otherwise would impede whatever momentum the organization may have achieved. The systems approach to high technology marketing does not restrict. In fact, it is easily demonstrated that it serves as an overlay to clarify and add a new dimension to vision and purpose for a marketing group.

An additional advantage of the systems diagram is to remind marketing management personnel that they carry a heavy responsibility with respect to the financial objectives for their company. Although the functions (usually the engineering and manufacturing groups) within the company which convert orders into sales are normally held responsible for gaining the desired margins, the cost–price work completed by marketing is for all practical purposes the baseline for the achievement of margin. It is essential for the success of any company that orders received are adequately priced so that performance against orders will yield the margins forecast. Cost studies are vital—they must reflect the best possible estimation of the cost of this conversion process. When they are high, orders are lost. When they are low, orders are won, but the profitability of the job suffers and often the profitability of the company. It is for this reason that the finance function must be included in any systems diagram. Proposals which are incorrectly priced have an undesirable impact throughout the organization.

One can argue that the blocks forming the system should be graded to show their level of importance; that is, their contribution to success. Such an analysis (to put numerical or some other value on each block) is an exercise in futility. Given such data, how would you use it? Would blocks receiving lower numerical values be bypassed in any later application of resources? Any block which is not properly addressed (by resources) will degrade overall company mission success. How much negative impact can you tolerate? Managers who operate their system successfully will tell you that it is easier to operate all portions of the system (a matter of degree in some instances) than it is to concentrate on one or more—leaving the balance to function on its own inertia or be dragged along by the stronger (more sensitive) issues. For example, this is precisely what happens to the manager who may, for one reason or another, elect to ignore the importance of proposal preparation techniques. A poor proposal can lose evaluation points and thus the job.

Some techniques work better for some companies than they do for others; some techniques I don't even discuss are employed by managers who have found a unique way to deal with the circumstances in which they find themselves. However, no matter what else you may learn, you will discover (or rediscover in many instances) that unless you attack your task from a systems point of view, I don't believe you can expect to maintain a precise *focus on objectives.*

A systems approach is nothing more than having the visionary bandwidth to see a total picture whenever decision nodes are encountered. The systems diagram should expand your current bandwith and thus improve management of the high technology marketing function.

CHAPTER FOUR

THE HIGH TECHNOLOGY MARKETING ORGANIZATION

The marketing staff for a high technology company can range from a single individual in a small organization to a full-time staff of hundreds in a multidivision, major aerospace enterprise. Such a staff may include marketing specialists, government relations experts, business data analysts, high level technical specialists, and personnel covering other administrative and nontechnical assignments.

Marketing managers have one major objective: to cover the marketplace with whatever disciplines are required to achieve the company mission. However, there is one facet of a marketing organization—one discipline—which is overlooked by many managers. It is the total company involvement in the marketing function—not as it appears on the company or departmental organization chart, but the manner in which the marketing (selling) task is shared by an entire company organization. The marketing manager who uses his total company organization to complete his selling responsibility is simply recognizing the positive impact this can have on the customer community.

This chapter will review organizational practices and describe how successful high technology marketing managers have organized their departments and employ the integrated company marketing concept.

ORGANIZATIONAL AND OPERATIONAL RELATIONSHIPS

Figure 4-1 shows the high technology marketing manager function as the hub of the company new business acquisition process. This responsibility is usually assigned to a single individual. However, the complexity of the task of acquiring new business cannot be visualized by looking at the size of the department or the functions the marketing staff have been assigned. Instead, one has to examine the total company organization, their operational philosophy, and their total business environment to grasp the true nature of the marketing thrust.

The "Customer" and "Competitor" identifications in Figure 4-1 need no further explanation at this point. These elements will be covered in greater detail in subsequent chapters.

However, I should stress here that marketing managers tend to allocate substantial percentages of their time toward the direction of the customer and less and less in the other directions noted in Figure 4-1. The customer *is* important; I do not want to mislead you. In my view, an inordinate amount of a company's selling task is structured around certain select customers. Perhaps that is a style born out of necessity; customers can often indirectly control a company through repetitive sole-source business. The company and the customer organizations can become so interwoven that it is difficult to separate one from the other. The point is that the marketing manager—the key individual in

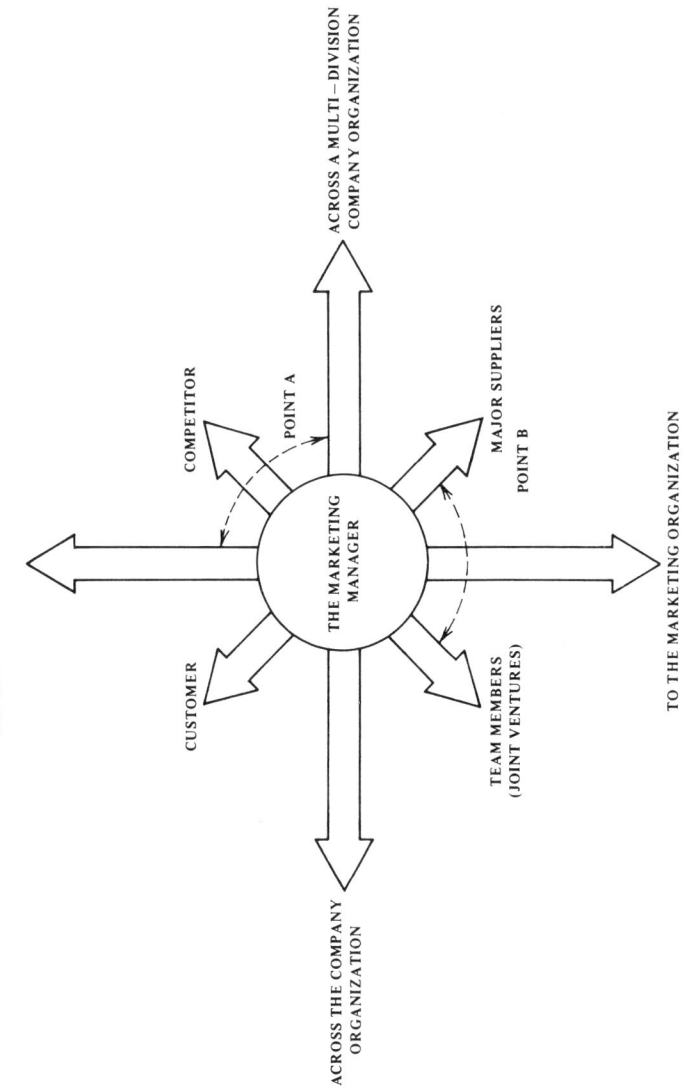

Figure 4-1. Marketing manager organizational relationships

the company's new business acquisition program—cannot effectively manage his functions unless he pays a certain amount of attention to the other elements portrayed in Figure 4-1 as well.

THE COMPANY MANAGEMENT HIERARCHY AND WORK PERFORMANCE

The company management hierarchy needs to be addressed here because of the major role played by all levels of company management in the acquisition of new business. Unless total company management becomes active in the new business acquisition process, the activities of the marketing manager and his staff will be so badly undermined that they will become ineffective. Most customer organizations have levels of management corresponding to the levels in your company organization, from your chief executive officer down through his senior organization. Almost every discipline covered by one or more senior-level managers in your company has a corresponding discipline somewhere in the customer organization. It makes no difference whether you are a prime contractor selling directly to some government agency or a second-, third-, or lower-tier subcontractor or supplier. Keeping these organizational relationships active is an essential part of the marketing (selling) process. Decisions to buy are as often made on the basis of the good working relationship developed between customer and company managers as they are on cost and technical evaluation. With cost and technical performance being essentially equal among competitors, the customer is most often inclined to buy from a company where the selling process has included the one-on-one contact between company and customer managers for each of the separate disciplines in both organizations. The marketing manager should use this relationship to his best advantage for every new business acquisition activity. Some programs may not require the heavy involvement of other company managers; however, the operational philosophy should be such that when it is required it can be exercised to the fullest extent.

Figure 4-1 shows the relationship of the company management hierarchy to the equivalent managerial levels throughout a multidivisional company (or corporate) organizational structure. This relationship is indicated by Point A. Companies within a group of corporate structure often have similar product offerings fulfilling needs for the same customer. This can prove to be a significant asset in the marketplace; it can also be a serious detriment under certain conditions. The marketing manager serving a company under this multidivisional organizational umbrella needs to fully understand the thrust of all other companies.

The marketing managers for each of the companies under a corporate structure need to develop a close working relationship. Because they may individually develop marketing thrusts involving the same customer agency, it is imperative that marketing planning information and customer contact schedules are carefully structured. Confusion at a customer agency (a procurement office) regarding who is selling what should be eliminated whenever such a condition arises for the first time. Even a joint presentation of capabilities and responsibilities for the companies under a corporate structure may be required to thoroughly clarify these organizational differences.

One additional intracorporate consideration must be fully explored by each of the representative marketing managers. It may not be unusual for one of the companies to encounter technical performance, cost, delivery, or quality problems on a contract while at the same time a second company may be actively pursuing a totally separate procurement—both with the same customer. The performance evaluators at the customer facility may look upon marginal performance at one company as indicative of the performance of the entire corporate organization. In most instances, this is an example of shortsightedness on the part of the evaluators, but it can become a significant factor in overall company evaluation when competitive procurement decisions are processed. There are no set rules for dealing with these conditions. However, each of the marketing managers must be realistic in representing his own company's performance and fully prepared to assist his counterparts within the corporate structure when the need arises. Severe problems at any single company in a multidivisional organization are usually addressed by higher levels of management. To do anything less will often degrade marketplace acceptance to such an extent that even the best individual marketing efforts for some new procurement will suffer or become suspect.

One of your best marketing tools is a satisfied customer community and thus a concerned and cautious competitor group. Let's explore this further. Many marketing managers will argue the value of such a relationship, but it is measurable and should form a part of your total plan. We have reviewed the impact of a single company failure in a multidivision organization. But what can you do when that failure is yours?

It is not unreasonable to assume that at some point your overall company performance against certain contractual requirements may be less than required. In fact, most companies in the high technology business area have suffered setbacks with respect to technical, schedule, and cost performance. It's the nature of the business. Many competitive procurements are won on the basis of optimistic structuring of technical and cost proposals where the inherent risks are considered good business risks. However, once the job is in the house, the plan for controlling the risk elements falls apart. Marketing cannot turn its back on the customer at this point; it must share with all company functions the responsibility for any performance shortfall.

The development of recovery plans for any project (contract) that has performance, schedule, and cost difficulties must also take into account the past relations with that customer as well as the relationship a company wishes to develop and maintain in the years ahead. This is why a marketing group must also contribute to the corrective action plans and presentations to customer evaluators. With performance on past and present contracts serving as a leading evaluation criterion for the award of additional work by that same customer, any company (and the company's marketing manager) will want to avoid any misleading or erroneous conceptions formed by the evaluators. The best course of action is to correct the shortfall as rapidly as possible, and thus protect any overall program plan held by the customer. Terminations can be costly indeed. The company's reputation suffers irreparable damage; the marketing manager's task suddenly reverts to one of defending himself and his company in the marketplace. Both can have a serious impact on the future of the company.

The alternative, of course, is to achieve most of the program milestones and other requirements precisely as defined in the original contract or as subsequently modified through careful negotiation. There simply is no substitute for performance. As the marketing manager, you must develop the ability to see the risks inherent in all new business acquisitions and then strive for risk reduction plans and programs that will ultimately satisfy customer requirements and protect the company's image in the marketplace. Some marketing managers see customer satisfaction (acceptable to outstanding performance on all current work efforts) as the most important factor in the total marketing effort. Many procurements are won solely on the basis of past performance. That, by itself, should serve as an adequate reminder.

There is one additional condition we need to explore with respect to multidivisional organizations. In Figure 4–1, I have indicated a relationship between the marketing manager and his team members or joint venture partners. This is usually nothing more than a formal agreement to work together in a synergistic mode to accomplish some goal (a new award) which either organization could not accomplish alone. It is fairly common practice for two or more companies in a multidivisional organization to form such teaming arrangements. A more general practice is for totally separate companies to enter into teaming or joint venture agreements. The marketing manager must again play a leading role in defining the purpose of the agreement and selecting the right partner. Selecting a team member from within a multidivisional structure offers some advantages. The contractual aspects are usually less complex. The treatment of company proprietary data becomes less cumbersome. In Chapter 9, I will review some of these teaming concepts as they relate to win strategy.

SUPPLIER RELATIONSHIPS

Continuing our discussion of the elements of Figure 4–1, a relationship is shown between the manager and the major suppliers to his company. Developing a close working relationship with a supplier's marketing and procurement organizations can be extremely beneficial. Many of your major suppliers also sell their products to your competitors and to other major contractors. They may have marketplace insight you need to more fully develop an overall marketing plan. Whether or not your suppliers will share all their marketplace information with you is a matter of conjecture on my part. However, the path to that information is open to you to exploit to the fullest extent possible. Furthermore, these same major supplier organizations could also serve as team members for your major new business acquisition plans. In Figure 4-1, this relationship is depicted at Point B. Gaining an exclusive teaming agreement with a supplier will drive your competitors to some other source. Developing mutual understanding and respect is a very valuable asset, particularly when the supplier has achieved recognition in the marketplace for his excellent performance.

RELATIONSHIPS WITHIN THE COMPANY

Next, let's examine more fully that part of Figure 4-1 relating to the company organization.

Total company involvement in the marketing process is not fully recognized at many organizations. While others may involve a substantial number of nonmarketing employees, they don't fully realize how vital that involvement is to the successful capture of new business opportunities. Successful marketing brings company and customer disciplines together to develop a mutual trust and understanding of their respective positions.

Company and customer technologists should meet to fully resolve technical issues. Program support personnel should meet to develop a precontract working relationship. Company management personnel should meet their high level counterparts within the customer organization. These meetings, whether formal or informal, are essential for the development of an evaluation baseline for both the company and the customer. For the company, to support the decision-making process regarding the pursuit of new business opportunities (and how); for the customer, to further support the study and evaluation during the source selection process. (Chapters 8 and 11 will address these issues in far more detail.)

Customer organizations acquiring a perception of your company only through routine marketer contacts cannot accrue a broad and accurate evaluation baseline. In fact, it has been my experience (and confirmed by most marketing specialists) that you must maintain a high level of company representation at your customer's facility whenever that customer is targeted by your new business acquisition plan.

You cannot expect members of the customer's organization to favor you in an evaluation process until they know and trust your company, your key personnel, and your methods of operation. This mutual trust and understanding may even have to extend to the highest ranking members of the organizations. It certainly includes marketing, contracts, procurement, engineering, and various support personnel when the scope of the business under consideration embraces all of these disciplines.

The favorite example of all experienced marketers is that company representatives should establish and maintain contact with the ultimate user of whatever commodity they sell. This is particularly true in the high technology business area where the customer is part of the Department of Defense or another government agency. It is essential for the designers of a new system to meet with the actual user—the person who will eventually operate the system by twisting the knob, turning the dial, pulling the trigger, or whatever.

Procurement documents (specifications) can only describe a requirement in terms of levels of performance under environmental and general usage conditions. However, these written instructions are not a substitute for the direct impressions obtained by an equipment designer when he, for example, drops into the interior of the army's new battle tank, sits in the pilot's seat in an F-18 or A-10, or even observes training exercises at Fort Bragg. Bringing that user viewpoint into the laboratory is probably one of the most difficult tasks for the designer. Successful marketing managers, however, have already learned how these visits can protect against an overrun, a schedule slippage, and a marginally acceptable design. It's just good business. And, the practice isn't reserved for companies doing business directly with the government. It is the commonsense business practice at any level of participation, including lower tier subcontractors and suppliers.

THE MARKETING ORGANIZATION

To begin with the development of the marketing organization itself, you should make some preliminary determinations with respect to how you want to represent products and product lines in the marketplace. The representation can be either product-oriented, customer-oriented, or ever a hybrid. We'll examine these organizational concepts.

Many marketing organizations I have studied show a single product or group of related products (a product line), represented by a single marketer (or an integrated subgroup within the major marketing function), to all of the company's current customers and all potential customers. There are some very good reasons why this is a practical organization. The marketer or marketers can become fully cognizant of current company posture with respect to these products and can also concentrate on them across all customer requirements. This provides for a better understanding of trends in the marketplace for a product family, a better opportunity to recognize needs developing in different sectors of the marketplace, and a much better opportunity to assess the strengths and weaknesses of competitors.

While there may be differing procurement practices across such a wide customer base, this is not regarded as a major deterrent to adequate representation in the marketplace. Differences in procurement practices among the various government agencies are not so diverse as to cause an experienced marketer much difficulty. And, when selling to government prime contractors or their subcontractors, each tier is subject to the same general government procurement practices. This tends to level out individual preferences at each of the companies. Their specific (and often special) procurement practices can be easily understood and are normally no reason for concern. If the product line involves both government and commercial customers, the procurement practices become somewhat more complex; however, once a marketer becomes familiar with the procurement practices of the government, commercial procurement practices may seem like child's play.

This type of marketing organization for representation in the marketplace has some other distinctions that should be noted. A product family may contain some unique technological features which require a marketer with a special technical background. These unique capabilities are essential if the product family is to be properly represented. Additionally, marketers having this unique technical background gain a broader acceptance within the customer community and are therefore better able to develop long-term working relationships with customer organizations. However, there may be one drawback to this method of representation.

Some customer segments may buy (have a need for) other portions of a company's product offering. When this occurs, the company marketing organization is usually forced into assigning two marketers to the same customer, each representing different product lines. While this may be confusing to the customer at the outset, the arrangement can be managed successfully if certain precautions are exercised. The customer should clearly understand why this kind of representation is required. This should not detract from the overall efficiency in dealing with that specific customer, provided this organizational preference for the company is announced in advance. Marketers sharing a common customer should make sure they are reading from the same script so as not to confuse the customer. The marketing manager must carefully control this shared-customer marketing concept. Often, when bringing a second company marketing representative into a customer's environment, even though different products and technologies are represented, the action can be upsetting to the customer and may even detract from some of the achievements made by the original marketer. In severe problem areas, it is wise for the marketing manager to visit the customer to pave the way for shared marketing representation.

Customer-oriented representation schemes are not as common, but there are many circumstances when this practice is the best of all other possible choices. Many companies will assign single representation of a specific product or product line to a single government agency or even a single and vital prime contractor. The reason is simple once the factors involved are studied and understood. The business volume (and thus the need for full-time representation) may be sufficient to justify the cost. The importance of the customer in future company business plans may be so great that all other considerations become insignificant. And, the customer may even demand full-time representation because of business volume, or because the assigned marketer performs a well-rounded, efficient, and believable representation.

This organizational concept usually requires one or more additional marketers to represent that product or product line to all other customers. Most often, the decision to split representation in this manner is initially determined by the order potential. When results are below expectations or plan, the manager then examines the representation alternatives available to him.

Other representation schemes may involve a single marketer handling all company products in a single customer environment, or across a broad range of customers. This is usually successful only when similar technologies are inherent in all products and when the marketer is experienced in handling multiple customers and products. Unless the assignment is carefully monitored, the marketer may devote his attention to only those opportunities promising immediate orders; thus, the longer range opportunities may not be addressed properly. Skimming the marketplace for today's orders—and neglecting the longer range possibilities—can seriously affect the business development aspects of marketing representation. Almost every experienced marketing manager has learned (often by virtue of his lost business file) that dealing in futures in the marketplace is every bit as important as booking today's order. These subjects will be addressed in later chapters when we review market analysis techniques and target selections.

For complex product lines, it is fairly common practice to

assign total product line responsibility to a single, highly qualified manager having marketing, technical, program management, and other skills. A single manager must be responsible for the total control and management of all product line activities. The reason has been stated over and over again by company marketing executives. Single-point control provides an overall perspective for the product line, and is not biased by a strong engineering department or other equally biased discipline. Someone must achieve a balanced thrust for such a product line. The work in progress is essential, maintenance of the technology base is essential, directivity and continuity in the marketplace are essential, and planning for and control of the total product line thrust must be carefully managed.

You should examine your own organization to determine precisely how products, product lines, and customer areas are assigned. You can argue that your present organization is adequate because results are good—orders are booked at or near budget levels. I suggest that one of the reasons a company fails to show a better growth trend (even when its technology base and product mix have many positive characteristics) is the reluctance to modify and improve existing marketplace thrust. Complacency is a deterrent to growth; a marketer's complacency expands with each new order credited to his account. But, change for the sake of change is not a solution. Change should be implemented because study reveals an opportunity to gain a greater market share. Marketplace representation for your company, your products, and your technology base should have an inherent flexibility. And, the marketing manager should use that flexibility to his best advantage.

In this era of increased government spending for high technology products (national defense, for example) and the oniomania at procurement offices, every marketing manager should structure his organization for maximum new business capture. It's simply a question of how soon you want to "join in the fun."

Other marketing (business development) organizational segments are usually a function of business size and the complexity of the product mix and customer environment. Larger companies often retain senior engineering personnel within the business development group to do the forward thinking with respect to research and development, analysis of technology trends and indicators in the marketplace, and evaluations of competing product offerings. In smaller companies, this type of work should also be done, but perhaps as a subsidiary assignment within the engineering organization.

Consequently, the problem of market data accumulation and analysis, forecasting, and business planning confronts the marketing manager at all businesses—large and small. How these functions are organized and staffed is dependent mostly upon how each company perceives its business and its overall plan for growth and profitability. When these functions are not staffed or are treated as fill-in functions for administrative assistants, the foundation for the future of the business will eventually crumble. High technology businesses cannot survive unless they plan for the future. If they don't, the competition wins.

Some companies which have been studied show a combined marketing (business development) and contract administration function. The latter includes control of contract terms and conditions, negotiations, and post-award contract management. This organizational concept has some merit for smaller companies with restricted or specialized product offerings. However, selecting a manager for such a combined organization usually results in reduced efficiency unless the overall manager acquires competent supervision for each of the two functions.

Finally, since marketing includes the total spectrum of product and technology representation in the marketplace (and the development of need in many instances), it is important to keep in mind the vital role of *marketing communications*. (Chapter 19 deals with this subject in detail.) It is a function within marketing which is often overlooked or shunted to a low priority category.

CHAPTER FIVE

DEFINING THE COMPANY TECHNOLOGY BASE

The term *high technology* is commonplace. The term is used by writers and commentators who report on significant scientific developments, which now seem to occur on an almost daily basis. But high technology serves no real purpose until the pragmatists describe its relationship to the marketplace.

The rapid transition of high technology into products—the practical application of technology emerging from laboratories at the government, industrial, and academic levels—becomes the challenge for the marketing function. The relationship of newer technology to an existing company technology base and product mix requires the fullest attention of both technologists and business development personnel.

Assessment of the technology in a company's existing product line is essential; user requirements tend to reflect the state-of-the-art and near-term fallout from the leading edge of new technology. Knowing where existing products fit with respect to the current practical application of new technology is an essential element of every company's strategic plan. The ability to accurately read trends and indicators with respect to the flow of high technology will often dictate the decision process regarding new product ventures.

The marketing manager plays a very important role in this assessment process. It is the marketing function (usually) which translates marketplace need into statements of characteristics upon which company technical personnel can then grade the company's technology base.

TECHNOLOGY BASE EVALUATION

A company's technology base is the level of technical achievement inherent in its current products (the old, new, and emerging), the levels of technical expertise throughout the organization, and the accessible technology it can draw upon when required to do so by the needs of the marketplace. However, I would caution you not to include in this measurement those technology base building blocks which have not passed through your technical organization.

Often, companies will claim (or attempt to claim) levels of expertise in certain technical disciplines; then, when it becomes a contractual requirement to perform within that area, a series of severe cost overruns and scheduling problems emerge. A company can oversubscribe to a set of technical requirements only to learn later, when the program is under way, that severe shortfall is being experienced. It is simply a question of knowing beforehand what your technology base limitations are and attacking the marketplace where you can utilize the base to its best advantage.

I believe it is essential for you to measure your technology base with respect to the new business acquisition targets you have selected and now have in your budget and forecasts. I would hope that target selection included the process, but for those who have not completed the measurement, let's explore the techniques by examining Figure 5–1.

TECHNOLOGY BASE EVALUATION

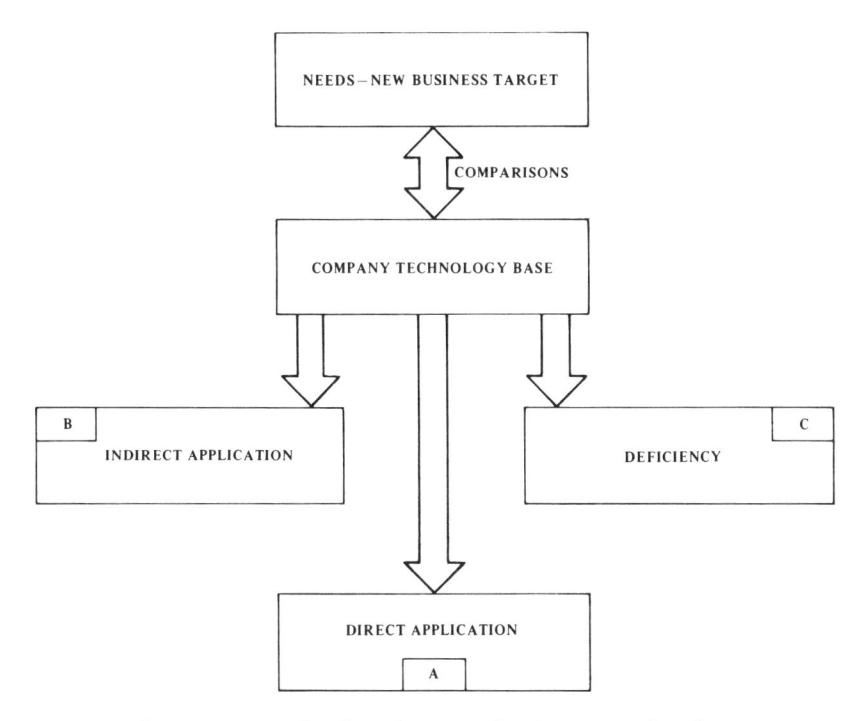

Figure 5-1. Technology base application to marketplace

For every new business target you are now pursuing, work the diagram requirements by dividing need into the three categories shown—A, B, and C. This will define the technology base. By assigning technology base strengths (capabilities) to your target programs, you are, for all practical purposes, defining the strengths of your technology base.

The listing of strengths and weaknesses for the base cannot be completed accurately unless some standard for measurement is set beforehand. I know of no better initial standard than the target programs in your budget. You may be surprised with the results. I assume, of course, that your analysis is truly objective. There will be a strong tendency to downplay the need to better fit an existing base. You probably wouldn't want to leave yourself open to criticism for selecting targets and allowing them to reach budget level where technology base deficiencies are noted, and you have no definitive plan to overcome these deficiencies.

The objective is simple, yet few marketing managers in my sample have an established and formal program for measuring their technology base against needs—their targets. Instead, they rely heavily on other, less reliable evaluations.

NEW BUSINESS TARGETS AND THE TECHNOLOGY BASE

There are many steps to follow in selecting new business acquisition targets, many of which will be discussed later. Of all the evaluations, none is more important than the relationships to your technology base. A company must identify risk associated with new business acquisition; the greatest risk is perhaps the assumption that somehow the technologists in the company laboratories will find solutions to all technical problems once the contract has been awarded. Often, the engineering labor content is so sadly understated that cost and schedule problems will result, or dollars are added to cover risk to push the cost proposal away from the competitive window. The best solution is to determine beforehand what your technology base can provide and what you need to acquire elsewhere.

An interesting aspect of technology base definition and evaluation is the multiple and sequential program contribution over some finite period of time. Companies devoting substantial resources to the continuing development of their technology base also seek complimentary targets (marketplace needs) to further build the base. If, for example, a future large and highly desirable Program C will require the near-term experience of Programs A and B, then the company strategy may be to win Programs A and B by virtue of whatever pricing adjustments may be required. Since *win* most often equates to *lowest price* (other evaluation criteria being essentially equal), a company investment favoring that objective is not unreasonable. However, such a strategy must be carefully controlled and based on the very best program evaluations possible. Investments in the technology base using this technique are a high risk at best. But, technology base expansion through program activity is quite often a better approach than direct application of investment funds for research and development to achieve the same objective. Performing Programs A and B (using our example) may have additional benefit beyond creating a win environment for Program C. Given the choice, I believe most companies would prefer to obtain almost all of their technology base enhancement through program activity. Unfortunately, that avenue is filled with considerable risk.

In Figure 5-1, one dimension is not shown but requires some explanation—time. Marketplace needs embrace today's requirements and the requirements expected to be established and defined for whatever future periods you use for planning purposes. Traditionally, companies plan three to five years in advance with respect to their technology base. Some large and aggressive companies with substantial resources look forward a decade or more. They help to define need for the user community. What they will have to sell—or promise to have—becomes the need.

Substantial government investment in research and development addresses the long-term need. An excellent example for high technology is the government's funding of industry activities related to VHSIC (Very High Speed Integrated Circuit) development. The companies involved are, to varying degrees, supplementing government funding with their own because they see the marketplace need developing. They have already described their technology base deficiencies related to that need. They are investing today for the future.

It is essential though that we understand the segment of technology addressed here. I do not include the research laboratory activities that deal in early concepts and experiments for whatever technology discipline you may wish to use as an example. There are hundreds of exciting examples of basic research now being conducted in government and industry laboratories and in the many outstanding research laboratories associated with colleges and universities. It is that continuing research that contributes so much to our world leadership role. However, that research fallout is not quite ready for application on a full-scale basis and, thus, is not of immediate concern to the marketing manager addressed by this text.

MAINTENANCE OF THE TECHNOLOGY BASE

Company research and development programs are not usually established to bootstrap the organizations, although any research and development task in the laboratories will raise the capability levels of the participants. Research and development programs established for the sole purpose of training technical staffs (providing them with the opportunity to raise their level of expertise to that of the competition) need to be carefully structured and controlled. Otherwise, funds will be forever expended to raise skill levels in disciplines which may have no relationship to the marketing thrust. However, marketing managers, having the responsibility to acquire new business, should not extirpate company research and development efforts by insisting that whatever is contemplated in this regard has already been done or has already been undertaken by others. A certain amount of hands-on effort is essential.

Engineering personnel will often argue for their research and development allocations without first having studied the status of the technology. Their emphasis is often associated with specific tasks that may appear to be a reiteration of something which has already been accomplished by others. The task has already been addressed, results are fairly well known, and applications of the new technology are beginning to appear in new products currently being developed by others. They still argue—even in the face of factual data—that they haven't done it in their own laboratories. They will argue they need the hands-on experience they can accrue only by having the project within their jurisdiction. They argue it is easier to write about a new technology in their next proposal if they have the experience of the laboratory environment. Here, the marketing manager can provide valuable guidance.

Technology, when not applied, is just technology. It provides fodder for the dozens of periodicals that serve to praise the technologists. But it does not provide new products, jobs for employees, and growth for your company. The tinkerers in the laboratories must be taught to relinquish their grasp long enough so that the pragmatic applications engineer can sort out the wheat from the chaff and adapt that technology to material for the marketplace. Technology has value only when it is applied. In my view, the greatest share of your R&D dollars must be spent on projects that will allow you to move out of the laboratory and into the marketplace on a timely basis. Each such R&D task must have a roadmap for that progress.

Much is written about the technology *freeze*. It is a subject addressed often by government system acquisition managers, as well as industrial managers involved in new product development.

TECHNOLOGY APPLICATIONS

When, in the program schedules for a system or product, it becomes prudent to enter full-scale development, the decision point is reached at that time to gather whatever proven technology is available (and applicable) and wait no longer for the promises of the R&D technologists. This does not curtail R&D activities; it only suggests that periodically one has to develop the new product for the market utilizing the available technology. Unfortunately, we often see product development cycles of such length that at the point in time in the future when the user (consumer) has the new product in his hands, the inherent technology is several orders of magnitude *behind* what the technologists can then offer.

There are no easy solutions. We all want the best that technology will provide. However, we must also give the pragmatists working with new technology an opportunity to complete their translation of that raw data into workable, usable, and cost-effective systems and products. We must all learn to use the current product (the interim solution) while we attempt to accelerate our often lengthy and costly product development cycles.

Military system planners have found no easy solution. They may never achieve anything but a minor improvement in isolated cases. One could visualize a situation where you bring a new and totally tested system or product into the inventory (the marketplace) every two or three years. This requires a continuing, accelerated, and probably totally unaffordable program plan. For many consumer products, this is routine. For complex military systems, it is out of the question.

Technology insertion is an attempt to keep a system or product current. Ideally, if one could insert new technology into a system by a simple module replacement, the threat of technology obsolescence for many products could be avoided or at least better controlled. To be valid, however, the developers of new systems and products must learn to save a place for new technology insertion in their designs. There again we can visualize a cooperative activity within a company where the key managers of research and development, product development, and marketing structure the best short- and long-term program.

The most recent concept to deal with problems of lengthy full-scale development cycles and the speed with which the progress of technology can overtake a program (a system or some product thereof) is that of *preplanned product improvement*. For some, perhaps, it is only a fancy term for technology insertion. However, the difference is embodied in the concept that the system procured today takes its bite out of the book of proven technology (ready for practical implementation) to obtain full-scale development completion and deployment in a shorter time frame. Then, as an integral part of the total program plan, milestones are established for inserting *planned* modifications. The milestones generally coincide with the consensus regarding when new technology maturity will be achieved. Technology insertion and preplanned product improvement differ only in that a schedule discipline is established for the latter when a system acquisition plan is formulated.

Whatever you care to call the concept, the simple fact is that you control technology insertion. You don't crowd new technology into a system or product until the pragmatists have used their required lead time to achieve cost-effective implementation criteria.

TECHNOLOGY BASE MANAGEMENT

Management of the technology base for a company is vital; the marketing manager must also contribute to this management activity. Decisions regarding technology application cannot ignore the marketplace; decisions regarding the further development of a company's technology base cannot ignore the marketplace. And so, marketing becomes business development. Technology is the basic framework; all else is part of the *fleshing-out* process.

There are many facets to the research and development side of your business. It is difficult to articulate fully in this limited text the kinds of R&D a company should pursue or even whether company investments in R&D contribute all that much to future business. Each company's management should study its specific programs and planning to determine specific needs. Given the opportunity to measure the effectiveness of R&D program activity at any randomly selected company, I believe I would find some mismatch between their current market thrust and R&D, between their business plan and R&D program, and between the strengths (or weaknesses) of their technology base and R&D programs. Your technology base assessment and programs for the maintenance and expansion of that base must be given full consideration during every phase of marketing and business planning. In any high technology business, to do otherwise will result in a deteriorating base and loss of capability to remain competitive.

Technology base management is not a new concept for the future. Successful companies have always managed their technology base—some better than others. Companies milking their cash cows while neglecting technology base investments will be caught short in the marketplace. The giants in industry who may have been guilty of such neglect in the past are busily engaged in aggressive acquisition programs to bring the technological advantages of others under their control. It's a viable concept for companies with acquisition muscle. Also, many are modifying their investment initiatives to reflect greater infusion of corporate and company assets into research and development. It is a baseline requirement for any company desiring a leadership role or even maintaining the one they now enjoy. Here, the marketing or business development function plays a leading part. For without the essential guidance of trends and indicators in the marketplace—the true assessment of marketplace need—technology base enhancement activities may be misdirected.

CHAPTER SIX

THE CUSTOMER

A chapter dealing exclusively with the customer may appear unnecessary to some. However, every high technology marketing manager—whether he will admit it or not—places the customer community on a par with his company in terms of the attention it receives. Variously described as the marketplace, the buyer, or the user, the customer should not appear as a distant entity that places orders, but rather as a full partner in the business endeavor. Without the close relations which will enable a full partnership to develop, current and future business objectives may not be achieved. Business relationships will take many forms and are developed for many reasons, but the broadest definition of customer relations must apply.

Customer identification and development is a major and vital segment of the marketing manager's daily routine. However, knowing who may buy what you have to sell (or may hope to sell in the future) is only an initial and indefinite step in the marketplace. The high technology marketing manager may have developed a good business base involving only a few select customer areas. Should he stop at that milepost? Can he expand his customer base? Is his expanding technology base addressing new customer needs as well as new customers. How do you address a broad customer environment with a limited budget and limited manpower? How many marketplace probes can a company successfully pursue? Are they the right ones—promising the best new business opportunities?

These are questions facing all marketing executives. Sooner or later they become the questions related to survival for the manager dealing with high technology in the marketplace, and thus survival and a growth pattern for the company.

Figure 3–1 in Chapter 3 depicts the range of customers available to the high technology business organization. Many of the largest businesses dealing in high technology products work almost exclusively with government agencies or prime contractors. Smaller businesses seldom serve as prime contractors to the government. Instead, they serve as feeder suppliers or first- or lower-tier subcontractors.

Export and nongovernment (direct consumer) needs are served by the majority of high technology firms. To a certain extent, export of high technology is controlled. However, there are many examples of our highest levels of technology being acquired by foreign governments and businesses. Their acquisition of the results of our ingenuity are not always associated with the products we are allowed to export.

The argument regarding the propriety of this practice continues. Spokesmen for the academic environment view the exchange of scientific data from a much different perspective than our national intelligence groups, who believe we give away too freely the fruits of our research. And, while our country is in what has been described as a *twilight zone* between war and peace, technology export control becomes a real issue. When also faced with economic conditions that cry for greater foreign sales—in any form—the problem is exacerbated beyond general comprehension for the technologists and the marketing managers addressed by this text.

CUSTOMER LISTINGS

Figure 6-1 provides one example of a format you may employ to better accumulate your customer base data. It may only serve to

BUSINESS CATERGORY	CUSTOMER IDENTIFICATIONS			
	INACTIVE	CURRENT	DEVELOPING	FUTURE

Figure 6-1. Customer identifications

provide an additional discipline within your marketing organization or some additional level of control over your general planning activities. For the most part, companies can readily identify their customers for each of the four categories shown. However, a listing is useful in a format that helps to initiate the evaluation and study process.

The need for listing inactive customers may not be fully understood, but may reveal some trends you have overlooked in the excitement of gathering new customers and pursuing the opportunities they have described for you. It is not unreasonable to ask that you determine precisely why any customer is listed in the inactive column. It may be as simple as some change in your product mix or technology base, your loss of interest in that share of the market represented by that customer, or a reflection of the competitive nature of the business where you must relinquish your share to the competitor with a cost–price advantage. But, for whatever reason he is shown to be inactive, you and that customer enjoyed a business relationship for some period of time. If the customer elects to procure from other sources because of your performance shortfall, that should be noted. Losing customers for this reason is a very serious matter. An inactive customer listing may only reflect the nature of your current business thrust and his procurement requirements. It would not be unusual for that customer to be listed as inactive while also appearing in columns identifying future customers. For example, you may not currently be selling to a certain Defense Department procuring office because previous procurements were completed and no new requirements for the same or similar materials have yet to be developed. However, your performance on all previous work was completely acceptable. Like good suppliers, good customers come around again and again.

The listing of current customers according to the format shown in Figure 6–1 may appear to some as another nonessential task. These are customers buying from you today. You have contracts with them to provide the products or services listed in the left-hand column. I suggest you list only major customers; customers identified with small equipment contracts, spare parts, and minor services should not be listed since they only clutter the chart and make later analysis more difficult.

Customers listed in the developing column are not yet under contract, but you are actively pursuing their procurements in the near-term. In this column, have you included an inactive customer? This would not be unusual. It means that you may still have a good relationship with an inactive customer and are addressing his new procurements. The developing column should have some length and substance. It tells you something about your near-term marketing thrust.

Then, in the right-hand column, list customer organizations you have identified for future business. These are organizations which are buying the products and services you offer but for which you do not have a fully developed marketing program.

CUSTOMER LISTING EVALUATIONS

Let's now examine the purpose of the listing, and cite examples which may provide a better understanding of the value of the chart as a tool for evaluating your marketing program.

Since the government is the largest single buyer of high technology products and services, it may logically follow that the marketing manager should survey the total complexity of this huge customer and plot a course for every buying segment. However, your listing would become unmanageable. If you are dealing in total systems—missiles, aircraft, ships, submarines, and the like—your customers (the government buying offices) are easily identified. Only a few offices procure these large and complex systems. You don't have to dig around in the marketplace to find the right buyer. However, this advantage is not available to the thousands of other business organizations desiring to do business with the government for the first time or wishing to expand their government customer base. Fortunately, the government generally tells you what it wants to buy and identifies the offices handling the procurements. The problem for the marketing manager and his staff is one of developing a sorting process—to match procurement plan to technology base—to identify the target agencies, programs, and procurement offices. The development of your customer base, then, is somewhat of a function of your current business size, the growth objectives for your company, and your budget for market development.

Many companies in the high technology product and service segment actively market only those agencies having identifiable procurements with near-term capture probabilities. Mostly, these are budgeted programs for the company; winning or losing the contract is often the difference between company success and failure, between profit and loss, and/or between growth and stagnation. But what about the future? Can a company continue to ignore all other agencies who may have requirements falling within the reach of the company's technology base? The answer is obvious, but the question of how to market this total buying environment is often cited as one of the most troublesome aspects of the marketing manager's job. Your shortfall is never more evident than when you learn that your competitor has been awarded a contract for some product or service and you and your marketing staff were not even aware that the agency involved had planned for the procurement. Every (or nearly every) planned procurement matching your technology base should be in your listing of target opportunities. Then, that customer belongs in the listing shown in Figure 6-1. Whether you participate or not is a matter to be decided later; the fact remains that at least your marketing activity had identified the opportunity. We will give this subject much broader consideration later. Here, we are simply looking for the methodology for identifying and expanding the customer base.

As you may have already discovered, I lean heavily toward the graphic or diagrammatic portrayal of a system or concept to better visualize scope, information flow, and interfaces. And so it is with customer identification. The objective is to draw you away from a mental image you may have of your business environment and toward a simplistic but effective layout you can use to further study your current posture and course for the future.

Your customer may be initially described as any user who buys or may reasonably be expected to buy your products and services in the future. For some, this identification may be so complex that management (control and development) becomes unrealistic. That is not the intent. Having already examined your technology base, you can begin to cut away at the customer listing to achieve the best possible entry points for new marketing thrusts. In a sense, it's part of the new business target evaluation process we will cover later. However, what we should strive for here is customer identification without the restriction of the specific and planned procurements you have in some target listing. Customer listings reflecting only your selected targets are too restrictive.

CUSTOMER BUSINESS PRACTICES

Having identified a customer base, you need to evaluate their procurement practices. How do they do business? What is the accepted routine for developing a business relationship with the organizations you have identified?

To begin, your government customer will be your most complex customer. He can be kind, generous, forgiving, and dependable. He can also be lazy, indifferent, incompetent, unrealistic, unsure of requirements and plans, and inaccessible. What he buys may be poorly described, inadequately funded, subject to change and cancellation, and so laced with rules and regulations that even the most skilled among us enter into a contractual obligation with trepidation. But, that is the nature of the beast. The vast majority of high technology firms are doing business with the government (either directly or indirectly) and most escape with their corporate hides and a modest profit. For all of the criticism levied on government methods and procedures related to the procurement systems it employs, one has to reserve some praise for the thousands of dedicated procurement personnel who somehow find a way to commit billions of dollars each fiscal year for every imaginable product and service.

For years, industry and government experts alike have reasoned that the government's procurement system needs a major overhaul. Unfortunately, the task is so complex that personnel assigned the responsibility of doing this never quite reach the levels of accomplishment promised by their objectives. The time period for completion of the task often exceeds their willingness to sustain the essential contributions. Then, new spokesmen arrive on the scene to argue that what has already been accomplished needs to be reviewed. Nevertheless, some progress is being achieved toward standardization of procurement methodology throughout the government. All critics of the current system should remember that while movement toward the objectives is slow at best, the government will remain the largest single buyer of high technology products and services. Suppliers to the government will remain suppliers; few, if any, walk away from this segment of the marketplace because of procurement methodology. It is fair to criticize. It is also fair to contribute whatever expertise we may have while the transition to improved regulations moves forward.

Marketing executives for high technology industries doing business with the government maintain a high awareness of government activities regarding the restructuring of procurement policy, regulations, and system standards. Business decisions regarding the long-term outlook of a company also need to be based on how the government agencies will conduct their business. And, it is not that difficult to keep abreast of the so-called *modernization* of the overall system. The Office of Management and Budget and the Office of Federal Procurement Policy (as well as the Department of Defense and the General Services Administration) publish a never-ending flow of information regarding their studies and implementation plans. The extent to which this information can be helpful to each marketing manager and his staff is something I cannot estimate. However, total ignorance of contemplated change may have some impact on business planning for the future.

CUSTOMER LISTING EXPANSION

The listings used in Figure 6-1 are not limited to government agencies. Each of your selected industry customers should also be shown in the proper category. However, it may not be as easy for you to identify all of the possible industry customers, since you may encounter difficulty determining what they may procure in the future. Prime contractors will come to mind first because their buying trends are generally the easiest to identify. Awards to government prime contractors are given much publicity, unless the award falls under a security restriction. Within a matter of just several months, you can develop a very substantial listing of contractors and their awards. Fortunately, such listings are available from several excellent sources. It is from these listings that you can develop a reasonably good identification of industry targets for your products and your technology base. For companies involved in high technology for a number of years, these industry identifications are firmly established. It is simply a matter of good business practice. The listings should be reserved to only those organizations which can reasonably be expected to have a need for (to procure) the products and services you can supply. We are matching technology base to the marketplace.

MANAGING THE CUSTOMER LISTING

An obvious conclusion at this point should relate to the value of the listing. I will now identify two simple and standard terms we use in our business: You must *categorize* and *prioritize* the listing. (Join me if you will; we *can* find a way to bury these terms.) Nonetheless, the objective is to arrange the listing in such a manner that you can easily identify and then concentrate on customer segments promising the best return for your marketing investment.

You may argue that listing is a waste of your valuable marketing assets; your staff has other and more important assignments to complete. Since the objective is to identify customer locations procuring products and services out of your technology base, let's select, as an example, just one customer category for evaluation—the government laboratory complex. To be even more restrictive, we can limit the evaluation to only those laboratories under the operational jurisdiction of the Department of Defense. I haven't taken the time to count them, but my first estimate is that about 100 laboratory locations can be identified. List them; then determine what technology area they address. Categorize them by that criterion. Now you can begin to accumulate laboratory locations dealing in your technology. Contact them. What is their charter? What is their budget? What do they buy? Are they addressing needs where you already have some level of expertise? Are procurements in the planning stage that you should review for possible addition to your target program listings? Should your technical personnel meet with their technical personnel? I could continue this discussion for an indefinite period of time; my point is that you can initiate a very important marketing discipline by performing the listing process. You are identifying your customer community—you are developing the customer base.

There is one segment of the government marketplace that should be addressed here only because to defer discussion would leave a major void in government agency identifications and programs receiving major portions of defense and other agency budget allocations. We have discussed general market considerations, and for many that is sufficient. But, the government annually procures billions of dollars worth of products and services of a classified nature. The products may be classified as well as their application once they become part of government inventory. A large segment of the high technology industry participates in these classified programs—some are identified and some are not. It is a function of the classification level assigned to the program. Often, procurements falling into this category are not advertised, companies qualified to participate are already identified, and the competitive nature of the business is not as severe. Sole-source procurements are not unusual. Companies who participate in this area are in some instances enslaved to the customer they serve; it is the very nature of the business. Their marketing efforts are conditioned for that marketplace, and while almost all of the general guidelines we discuss in this text will apply, there are some notable differences. When our national security is involved, a free-wheeling marketing activity is not evident to the casual observer. However, within the community the participants play out their roles as vigorously as their counterparts in unclassified business areas.

A good customer base can be developed through both *active* and *passive* marketing techniques. Active marketing is defined here as the business development practice that brings you and your selected customer agencies together through the direct contact of company marketing specialists and customer agency personnel. You actively pursue selected targets that have become budget line items. But, for every budget line item you actively pursue—which, by the way, often defines your true customer base—there are dozens of other opportunities with other agencies having a close relationship to your technology base. You can't ignore them since they may represent an opportunity for the future. One easy solution practiced by most marketing groups is an energetic marketing communications program. It is this passive marketing effort that will keep your message out in the marketplace, stir the interest of procurement offices, and possibly lead you to marketplace areas you had not entered previously. All of Chapter 20 is devoted to this subject because I have learned that it is an absolutely essential part of any company's future growth plan. An expanding customer base is the essential element of any growth plan.

CHAPTER SEVEN

COMPETITOR ANALYSIS

The development of overall company strategy, as well as a specific strategy leading to the capture of an identified new business opportunity, is a primary ingredient of the formula for company success. Competitor analysis is also a critical ingredient. While most high technology marketing managers can name their leading competitors, very few can identify the new competitors appearing on the horizon or how they may totally alter a win strategy for the very next competitive procurement.

We will now provide a series of guidelines for competitor identification, analysis, and ranking, and the key trends and indicators that will most often define competitor business strategy and how the manager should use this data in his decision process. Methods employed to acquire competitor data are varied; but within the limits of ethical business practice, a sufficient data base can be developed. Much of the required information can be accumulated from what a competitor chooses to tell you about his organization, operations, and plans for the future.

I believe that competitor identification is as important as customer identification, discussed in the previous chapter. Competitor identification may also be as important as determining the strengths and weaknesses of your technology base. Why? When you venture into the wilderness of the marketplace, your competitors will be there. They will be contacting your customers. They will be proposing the same jobs you propose. They will win. You will win. However, one of the methods you can employ to win a greater share of the market you are pursuing is to implement a competitor analysis function within your marketing organization.

DEFINING THE COMPETITOR AND COMPILING A DOSSIER

I believe it is important to understand at the outset what is meant by the word *competitor*. True, the competitor is your rival in the marketplace. But, simply because his product line is similar doesn't necessarily make him your competitor. He becomes your competitor when he captures a sizeable share of the same market you address, when he actively pursues the same customers and their procurements that you have targeted, and when he beats you in a head-to-head confrontation for the target you both have budgeted. Consequently, you should now accumulate (as rapidly as possible) every bit of available intelligence about his operation. It is at this point that I want you to fully understand how I feel about intelligence gathering relating to your competition.

Many of the details of your business are not divulged to the general public for the simple reason that to do so would undermine your competitive posture. And so it is with the competitor. He also has proprietary data he must protect. I believe that except for isolated cases, companies do not actively seek to obtain their competitors' proprietary data and would not become involved in any activity that would compromise this data. It is the very nature of our industry and generally accepted behavior. Company reputations and professional careers are at stake; to compromise one's company or one's position is totally unacceptable. There is an alternative. Acquire a dossier on your competitor from the information he and his customers release to the general public on a more or less routine basis. I'll give you some good examples.

Acquire his annual report. Collect his product literature. Visit his booth at trade shows. Attend seminars where his personnel present technical and other papers. Buy and test his equipment. Evaluate his product performance through discussions with user personnel responsible for operation and maintenance. Acquire *sanitized*[*] government test reports. (However, clearly indicate the nature of the request and do not accept offers of report copies that contain proprietary information.) Buy his drawing sets from the government when a follow-on production procurement is scheduled. (This is not a commitment to bid.) Read his advertising copy. Attend the same bidder's briefings; listen for his questions. (It is often suggested that competitors will *plant* questions at a bidder's briefing to mislead all other competitors. Don't believe it.) Hire some of his key personnel when they respond to your recruitment calls. Use the Freedom of Information Act—but use it wisely. Maintain a complete listing of his contract awards. (Classified awards may not be generally released as public information.) Maintain a listing of jobs he was not awarded, and try to determine through general information channels why he did not win. Was it price, schedule, and/or technical performance shortfall? There are many other sources of information I could list; I believe the preceding examples will lead you in the right direction.

I will later discuss win strategy in a competitive environment. One of the key data items for the development of a win strategy is the competitor's dossier. Your objective is to determine from the data you have compiled how your competitor will approach a

---

[*] The *Department of Defense Dictionary of Military and Associated Terms* states that to 'sanitize' one must revise a report or other document in such a fashion as to prevent identification of sources, or of the actual persons and places with which it is concerned, or of the means by which it was acquired. This usually involves deletion or substitution of names and other key details. It is essential to note here that the term 'sanitize' has also been employed by many to denote that classified portions of a government classified document have been deleted, thus converting the document to an unclassified status. Declassifying a classified document by cutting out or obliterating those portions identified as classified—or believed to be classified—can be a dangerous practice. Declassification should be attempted only by the originator of the document and the person or persons responsible for its classification.

specific procurement. You can never be absolutely sure of his win strategy, but having the dossier will at least give you an opportunity to speculate. Many companies and their marketing organizations have achieved an excellent record with respect to slotting their competitors in a competitive procurement. Competitor track records over a period of years generally serve as a solid base for the evaluation and slotting process.

Strange as it may seem, the government departments and agencies dealing in high technology products and services are formidable competitors for specific work elements. The Department of Defense, for example, maintains an elaborate laboratory structure. Much of the work performed within the laboratories can also be accomplished by industry. But there is general acceptance of the government's argument that a certain *core capability* must be retained; otherwise, it would ultimately become totally dependent upon industry (and the academic community) for its technology-related decision processes. Procedures have been implemented over recent years to provide a more equitable division between work assignments to the government laboratory, to industry (in the form of contract awards), and to university-controlled research and development facilities.

The competition from foreign industry—much of it subsidized by its own government—is an additional competitor element that must not be overlooked. Companies dealing in the export market have learned that in-country competition is formidable for certain products and services. High technology products offered for sale in the export market (where approval from the U.S. government has been obtained) are usually not faced with such a severe competitive alignment in terms of technological content and product performance. However, many U.S. companies find that competitive pricing windows are much lower.

Foreign offerings in this country are not to be taken lightly. High technology product performance levels for foreign-built equipment may be as good or even better than our own. And, these competitors find a way to achieve the best cost–price posture for their products for many U.S. government and industry procurements. Furthermore, with foreign competitors allowed to compete against U.S. industry for lucrative defense and other contracts, marketing managers are faced with a new set of competitor evaluation problems.

This is not a proper forum for a discussion of the merits of the case from any viewpoint. But, there may be specific and excellent examples when the Buy America provisions should not be invoked. One needs to fully explore the international political advantage (assuming one has been established) while weighing the short- and long-term impact on U.S. companies and their employees. I personally believe that whatever political advantage we accrue is short-lived at best.

COMPETITOR EVALUATION

Competitor evaluation is essential. You can argue that the task is too complex for you to tackle and that you don't have time to accumulate competitor data files. First, it's not a complex task; second, it's a continuing and on-going assignment for any efficient marketer; third, you probably don't have that many viable competitors; and fourth, a data file system can be established for about the cost of a single file drawer and a box or two of hanging file folders. Make competitor data acquisition a standard discipline for your marketing and technical groups. Every bit of information accumulated should be added to the file in a logical sequence. The information is then available when you need it.

Files such as these have a tendency to gather unsubstantiated information. Portions of these files may also become obsolete in a short period of time. If you believe competitor files will be useful to you, then I suggest you establish a discipline for their maintenance. Substantiated data should always be included; superseded data should be discarded.

The indirect competitor category is often overlooked. However, these organizations can and do inject some degree of confusion into the evaluation process for new government procurements. I classify these competitors as indirect because they have no intention of submitting a proposal. Instead, they tend to confuse the proposal strategies of many of the viable competitors by simply appearing on the bidder's list. We will review this in greater detail in Chapter 9.

For the most part, these indirect competitors are practicing what many consider good marketing tactics. They enter the procurement cycle for an advertised government procurement not because they intend to respond with a proposal, but rather to study the procurement documents (including the specifications) for the purpose of better aligning their marketing efforts for the future. Just as often though, certain organizations appear on bidders' lists to receive the government's procurement package only because departmental procedure requires an administrator to order every available bid set for certain procurement categories. This shows a great lack of discipline within a marketing organization.

Most major companies have already identified their direct competitors. What is important here is that you also remember to focus periodically on indirect competitor listings. This will enable you to determine if one or more may be developing a real strength as a future competitor for some segment of your product offering.

Competitors can become customers; customers can become competitors. In high technology business—as well as in many other types of business—supplier and buyer roles are often interchanged. The decision to serve as a supplier to one of your major competitors should be based on a full evaluation of past, present, and future marketplace considerations. Disclosure of your cost and technology data to a potentially antagonistic competitor may be more damaging to your future marketplace thrust than the loss of an order.

Teaming and joint venture relationships between competitors are not uncommon in high technology government business. In fact, on many large systems the government actually encourages such team relationships. Government procurement specialists— knowing beforehand many of the strengths and weaknesses of the viable competitors—envision a more responsive, cost-effective proposal submittal and improved performance under contract when two or more companies combine their many resources.

Within industry, the win strategy of the competitors will often include synergism to obtain the highest evaluation by the procuring agency and thus the award. And so it is that development of a complete competitor dossier will serve a multiple purpose: (1) determining how the competitor may address a specific procurement, (2) determining how the competitor may be able to serve as your team member when a new procurement requirement suggests that relationship, (3) determining under what conditions you would consider serving as a supplier to him, and (4) aiding in the selection of your target (budget) programs and in the development of your win strategy.

CHAPTER EIGHT

HIGH TECHNOLOGY MARKET ANALYSIS

Market analysis means many things to many marketing managers. For the high technology marketing manager, market analysis can be defined somewhat loosely as the process employed to size the market for whatever products and services his company now provides or expects to provide in the future. But the analysis has many more facets. What percentage (or segment) of the general market addressed is available because of prior participation, customer acceptance, technology base, and product mix? What percentage of the available market can be captured? Is it expanding or is it narrowing? Is a new technology insertion required? Are new market segments becoming available to the company because of technology base expansion through research and development? These and dozens of other considerations confront the marketing manager as he addresses his current-year capture requirements and those he has identified for three, five, or even ten years into the future.

As noted in previous chapters, this text does not specifically address consumer products, and consequently in this chapter associated market research guidelines. But, as a high technology marketing manager for products predominantly used in nonconsumer products, you may learn a great deal by examining some of the techniques employed to define, size, and forecast markets for typical consumer products. The disciplines involved may be simple or complex; studying case histories will give you an appreciation of this important aspect of business planning for other types of organizations. At the very least, it should convince you that market research (or analysis, as we will define it here) is an essential part of your job.

A DEFINITION

The term *market analysis* confuses many high technology business development managers. I believe much of the confusion stems from the obvious (but erroneous) relationship to market research. Market research deals with the accumulation of a data base portraying the detailed characteristics of your customers (current, and those to be developed) and competitors. What has been discussed in earlier chapters is partially market research; however, I have added the discipline of developing a full understanding of your current and projected technology base. Without the knowledge of the technology base attributes and the potential for expansion, your market analysis will tend to be superficial and meaningless.

Up to this point you have studied your company's technology base to again identify the products you have to sell, to determine the status of the technology inherent in the products, to evaluate how well the products reflect the current state of the art, and to ascertain what you believe needs to be accomplished to upgrade the products to address current and future markets. You have also examined your current customer base (in terms of existing customers and what they traditionally procure) and potential customers (who buy products within your product family but from other suppliers). You have also examined your competitors. Given this vast assortment of data, how does one make use of it to the advantage of the company?

The next step is simply one of bringing all of this information into focus with respect to the total market in order to determine current posture. Out of this will emerge some very constructive fallout which, when used properly, should point you and your company in a new and growth-oriented direction.

The real value in doing this is not to fill in blank spaces with totals and percentages. Instead, it is having done the research that forms the baseline for the final computation. The summation may be essential in order to satisfy the monitors who need proof that the research was accomplished. The value to you is the insight you gain into the complexity of the marketplace, the elements of its make-up, the details of the segment your business addresses, and the confirmation you need to establish your basic marketing priorities.

At this point, it is important to understand that we are addressing a closed-loop functional series consisting of four major elements: (1) technology base, (2) the customer, (3) the competitor, and (4) the analysis of the market. You may begin to see that any analysis of the market for high technology products cannot begin until you have a full understanding of the first three elements, and that market analysis activities which tend to subordinate or even ignore any one of these elements will probably not be successful. One highly respected business development manager for a high technology company has the graphic shown in Figure 8–1 posted in a prominent place in his office. Study the illustration for a moment. You will see the key relationships you need to consider. This illustration (or any similar picture you wish to construct) may be a useful reminder as you address day-to-day issues. Ignoring any key element for very long will distort your view of the marketplace and your relative position.

Now, with these tools sharpened and ready for use, we can begin to address the marketplace. The key from this point forward is to remember the interfaces shown in Figure 8–1.

THE TOTAL MARKET

The initial task in the development of the market analysis activity is the recognition and sizing of the *total market*. This is perhaps

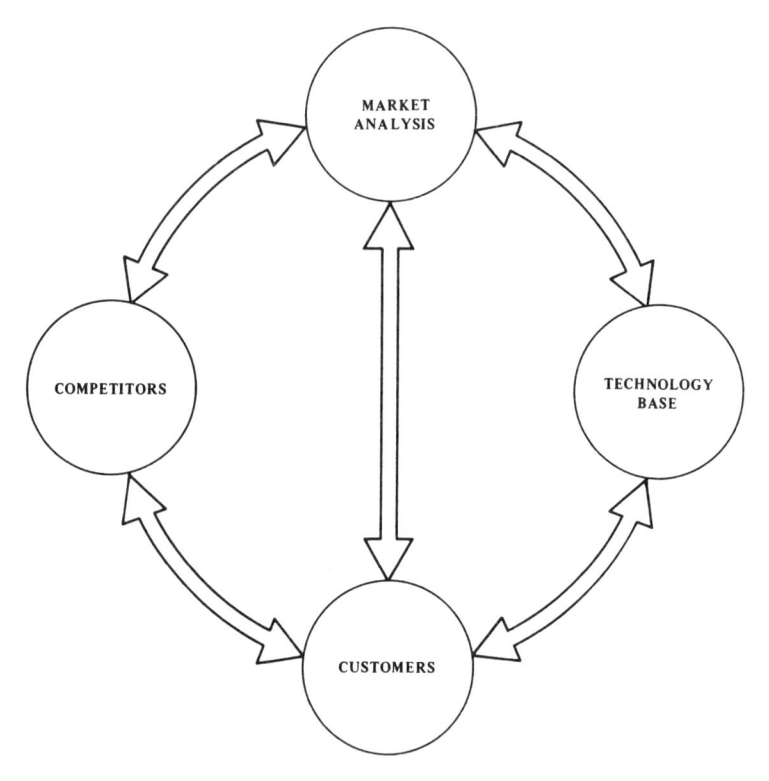

Figure 8-1. Market analysis relationships

the most difficult of the various tasks associated with market analysis. If you are a military aircraft manufacturer, your total market can be determined (for some defined period of time) to within a few dozen aircraft for both domestic and foreign markets. The Department of Defense will tell you how many they plan to buy over a period of several years, Congress will tell you how many they'll fund (on an annual basis), and, collectively, the executive and legislative branches of government (and others) will probably provide some indication of export potential. That's total market.

If you are the manufacturer of one or more of the hundreds of component parts of these military aircraft, you can develop a total market for your product(s) by using the total number of aircraft to be procured over some increment of time. To that you would add spares *and* any other applications for the same or a similar product. It is possible that the component can also be used in missiles, spacecraft, ships, tanks, and other military weapon systems. Count them; quantities are readily available. Also, examine nonmilitary procurements. Any component for military aircraft may have an application in commercial and business aircraft as well. Examine the export market potential for your component.

Remember that total market is the sizing of the marketplace for your component without regard to customer base, competitor share of that market, the specification for that component, and the position of your component with respect to stated requirements. The question you are trying to answer here is how many (of your components) are being procured by any customer over any selected increment of time.

For many manufacturers, the sizing of the total market may not be as simple as in our example. The question you have to ask is how many of the types of components you produce are being procured for each increment of time selected.

Many of the larger high technology aerospace organizations maintain market analysts. They study market trends (buying trends for both domestic and foreign markets) to form a basis for all other analysis work they will do. Smaller organizations, which cannot afford full-time market analysis personnel, can use several excellent market data service organizations which will provide total and segregated market studies. Frost & Sullivan and DMS, Inc.,[*] are two such organizations which have excellent market surveys (studies) available. It is usually more economical to procure outside service than it is to develop an in-house capability.

Total market determinations are vulnerable. The next step is to conduct a vulnerability study to determine the *hard* and *soft* increments. Gaining an appreciation of the factors causing a total market to expand or shrink (dollars or numbers of units) will provide some insulation. For an accurate analysis of the total

---

[*]Frost & Sullivan, Inc., 106 Fulton Street, New York, NY 10038; DMS, Inc., 100 Northfield Street, Greenwich, CT 06830.

market, you should consider the vulnerability aspects, quantify them, and establish correction criteria.

To continue with our earlier example, the Department of Defense (short of a declaration of war) will probably not procure as many new aircraft as originally planned or even funded. Traditionally, the quantities showing up in original government procurement planning documents are higher than the numbers actually procured over specified increments of time. Inflation is an obvious factor, but this insidious planning cancer aside, the total numbers are usually reduced by other budgeting actions, the conservatism that invades all major procurement actions at one time or another, and the inability to reach a consensus in either the executive or legislative branches of the government regarding major procurements by the military.

Quantity adjustments are often mandated because new technology offers—or promises to offer—an improved product at an earlier date. Using our aircraft example again, one has only to review the chaotic procurement planning within the Department of Defense and the Congress because of the promise of *stealth technology*—an aircraft design technology promising to render our aircraft less visible to current enemy radar and other systems. The tendency is to reach for the new technology and consequently the new product—often before it is ready for the market, the user, and the purpose for which it is intended.

Total market is defined by the following formula: *total perceived market* less the sum of all factors contributing to a reduction of the perceived market equals the *total real market*. Business planners will almost always size the market on the conservative side. If the eventual market ever exceeds original forecasts, the problem can usually be accommodated within the business structure. Conversely, major downward adjustments spread throughout the business and impact on manpower levels, capital investment, overhead structure, and other elements. Realism is essential for good business planning; applying good judgment with regard to the total market (the real market) provides a much better base for operation of the business and all short- and long-term planning.

Marketing managers have a general fear of using raw numbers out of the marketplace. Downward adjustments are usually arbitrary, ranging from 5% to as much as 30% reduction of the perceived market. The *a priori* reductions may be acceptable for many businesses. However, I would suggest that sufficient data are available to complete an analysis that will withstand most critical evaluations by higher-level business planners in your organization.

THE AVAILABLE MARKET

The next step in market analysis is the determination of the percentage of the total real market available to you. This is an estimate of the segment of the total market you could serve (not that portion you actually serve) based on current products and their technology. This requires careful study. Market segments may become available at a later date as a result of current-year research and development activities. Also, you may be inserting new technology into your current products, or developing spin-off products within the same family that will improve your capability to compete on a broader scale. We are attempting to determine the *size* of the available market—the market for your *product family*. Then, the same adjustment criteria you employed to obtain total real market should be applied again to establish a true available market. Now we can begin to think in terms of actual market size. Some business planners for high technology companies believe that market analysis up to this point is of little value; the next steps pull you down from the clouds.

THE SERVED MARKET

Served market is the percentage of the available market you actually serve—the segment of the available market addressed by your selling organization. If your computations show you are serving 10% of the available market, then you have essentially released the other 90% to your competitors and have probably established a *no-contest* situation in their minds—and in the minds of the customers.

This segment of your market analysis should indicate clearly where marketing aggression is required. It should point out that to increase your served market percentage you should increase the activities of your selling organization and/or introduce new products (or improved existing products). One final step is to compute market share; that is, what percentage of the served market are you actually converting to booked orders. Here we enter the realm of competitiveness. In the high technology business sector, competitiveness does not always equate to lowest cost. Buying organizations will often overlook higher cost to obtain a more suitable product.

MARKET ANALYSIS VALUES

One vital piece of information which emerges from a market analysis is the position you hold with respect to the technology inherent in your current products. A portion of the total market may be considered *firm* (over some defined period of time) because your products reflect technology utilization that will not render them obsolete in the near term; that is, your current products fulfill the stated requirement. Other products may be on the threshold of obsolescence, or may have been overtaken by the competitor's thrust to incorporate new technology into his own products. As your market analysis continues, you will begin to see the important trends—what new technology is required and how your current products equate to that requirement. Decisions are then required to either abdicate a specific share of the market (because of product obsolescence) or to organize and fund the research and development necessary to retain and possibly increase your total market share.

Many managers have told me that customers (government or industry) will define the market for you. All you have to do is resolve their buying trends into numbers of future orders for your products. This may be useful input, but it is superficial to the problem at hand. High technology is a dynamic term in itself; translated into the language of the marketplace it becomes complex, misunderstood, and even unpredictable. The reason is simple. Your products—the hardware items coming off the production line—will usually enjoy customer acceptance (and thus future orders) only for the period of time in which they retain their *essential high technology content*. Obsolescence can occur rapidly. Therefore, you must develop the ability to visualize trends in the marketplace—*the impact of the introduction of new technology*. You must also determine if the introduction of new technology is your current thrust or that of one or more of your competitors.

To sum up, market analysis may never become a discipline in your organization. You may be satisfied serving your current market—believing that your customer base and recent-year selling successes will protect you over the long term. Technology will eventually force you to reconsider. New technology evolves into improved product functions and utilization. It simply becomes a question of whether you move your business forward at the pace of new technology or remain a standpatter and watch your market share diminish.

CHAPTER NINE

TARGET SELECTION AND WIN STRATEGY

In the terminology of the high technology marketing manager, *target selection* is the process of identifying the new business opportunities (programs) having a relation to the company technology base and product mix, a near- and long-term profitability, and an acceptable level of win probability. *Win strategy* is simply the total company plan to acquire these selected new business opportunities.

On the surface, this may be construed as a straightforward procedure for achieving whatever growth plan a company wishes to project. But, for most marketing managers, the process is not straightforward. It is a serious and time-consuming process that builds and strengthens the baseline for the later achievement of almost all company objectives.

A versatile and energetic marketing staff can usually provide listings of dozens—and perhaps even hundreds—of new business opportunities. From that offering, the budget line item programs must be selected. How do you determine that certain programs are more desirable than others? How are program assessment criteria developed and then tested against a listing of new opportunities? Why is it possible to select specific targets (program opportunities) you cannot win? These and other critical questions are debated on a continuing basis throughout high technology industry.

The process can be organized to provide visibility the manager will need when final recommendations are prepared for submittal. Here, we will evaluate a process that can provide improved selection and thus a higher probability of a win.

Win strategy is then applied to each selected program. While many elements of a total company win strategy apply to all new program selections, a tailoring process is essential. No two procurements are exactly alike in structure, timing, and value. Tailoring a general win strategy to each specific program is a major step forward in the acquisition of that program.

THE COMPANY–CUSTOMER RELATIONSHIP

There are two operating systems involved in high technology marketing. The first system is contained within the user (the customer) community while the second is within your own company. This is portrayed in Figure 9–1. The two systems are interconnected by the interface subsystem which consists of dialogue (the transfer of information) and the products and services offered by your company. How well the interface functions has much to do with the success of your organization.

The interface module is shown in this illustration to suggest that some formal formatting of the exchange is essential. Some will suggest that the two systems are interlocked without the

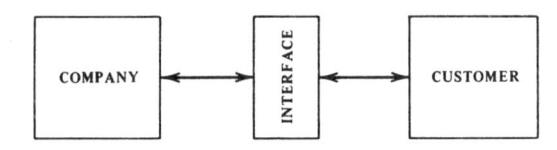

Figure 9-1. Company and customer systems and the interface

interface; however, I don't subscribe to that idea. The reason is simple. You must translate the language of the marketplace into the language of your company. User needs are usually not expressed in terms that will allow you to directly apply them to your technology base and product line. A certain amount of *curve fitting* is essential. Likewise, the products you offer may not satisfy user needs in all respects; some tailoring is usually required. Hence, the need for the translation subsystem. The key to company success then is the efficient interoperability of these two systems—determined mostly by how well the interface subsystem has been configured and how well it functions.

Figure 9-2 shows the relationship of each of the key elements making up the closed-loop system of *need expression* and *need fulfillment* in the area of high technology products. When you visualize the conditions of need and supply (need expression and fulfillment) in these terms, it will help you understand the relationships and perhaps better manage the elements under which you operate. For example, it does no good to express a need that cannot be satisfied by the technology. We hear of need pulling at some elements of the technology. We hear of technology creating need because of its sudden translation to maturity. However, need usually cannot be fulfilled in the near term unless technology is emerging from the laboratory that can be translated into practical use. Our government said: "Develop the atomic bomb." Years later, the terrible destruction we had conceived and brought to practical use was amply demonstrated. President Kennedy challenged government and industry to place a man on the moon. Ten years later the objective was achieved. These were needs expressed when technology was not yet ready.

The manager must carefully evaluate where his company technology level stands with respect to user needs and the competitive world around him. He must then decide what needs to be done to achieve a technological plateau that will maintain his competitiveness and provide products and services fulfilling marketplace needs. The system depicted in Figure 9-2 must constantly rotate and function. For some products the cycle is 3-5 years; for others, it may by 20-30 years.

The point of this discussion is simple—the dialogue between

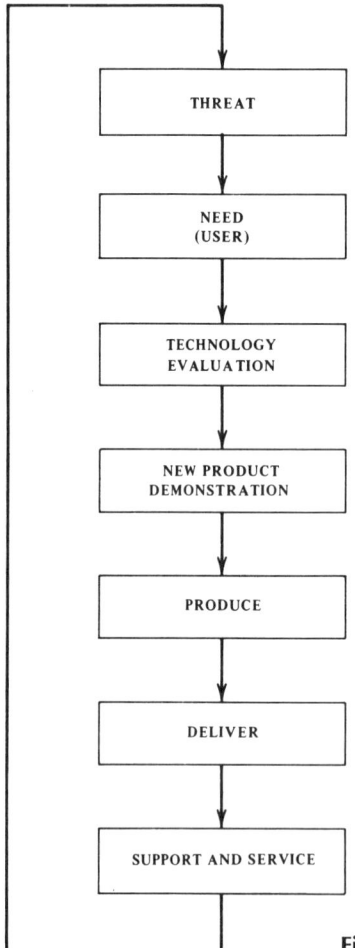

```
         ┌──────────────────┐
         │      THREAT      │
         └──────────────────┘
                  │
                  ▼
         ┌──────────────────┐
         │      NEED        │
         │     (USER)       │
         └──────────────────┘
                  │
                  ▼
         ┌──────────────────┐
         │   TECHNOLOGY     │
         │   EVALUATION     │
         └──────────────────┘
                  │
                  ▼
         ┌──────────────────┐
         │   NEW PRODUCT    │
         │  DEMONSTRATION   │
         └──────────────────┘
                  │
                  ▼
         ┌──────────────────┐
         │     PRODUCE      │
         └──────────────────┘
                  │
                  ▼
         ┌──────────────────┐
         │     DELIVER      │
         └──────────────────┘
                  │
                  ▼
         ┌──────────────────┐
         │SUPPORT AND SERVICE│
         └──────────────────┘
```

Figure 9-2. Product life cycles: a closed loop

your company and the customer community must be maintained at all times. This is essential if the direction in which the company is headed is to be corrected periodically to dovetail precisely with the needs of the marketplace. Dialogue provides visibility regarding the developing needs in the marketplace. But, you can help to develop the need by demonstrating to your customer the proper direction for him. Needs aren't always created out of the customer environment; they are just as often created by an articulate and energetic marketing effort.

Affordability is one of the key factors leading to the development of a need in the marketplace. When you can show your customer that, over the long term, he can accrue a savings by dismantling his current machine and replacing it with your new machine, you are forcing the need upon him. He may not be able to visualize his total cost of ownership. But, with an energetic marketing effort, you may be able to demonstrate to him that the change has *economic benefit*.

The only sure way to accomplish this part of your marketing effort is to visit the user community. This entails devoting considerable time and effort to learn what the customer has in inventory, how he employs it, how much it costs to operate, what its limitations are, and where improvements can be incorporated. This is business development.

In Figure 9–2, the block above the "Need (User)" block may require further indentification. In defense-related program activities, *threat* more often than not will dictate need. Our national intelligence groups continuously evaluate threats to our national security (both internal and external), and in so doing establish some of the basic requirements for new systems, products, and methods for dealing with threat. As new technology becomes available to our potential adversaries and they translate that technology into new weapons and improvements to their existing arsenals, we in turn see needs developing to counter this new threat.

Many industrial organizations work closely with the Department of Defense and other government departments and agencies in the areas of threat analysis, assessment, and evaluation. In this manner, they are better able to visualize trends in the marketplace with respect to changing needs. This early indication of need-change benefits these industry organizations in their planning for the future. For other industrial organizations, briefings are scheduled periodically to spread the early-warning data base. Thus, most of the key defense contractors and their subcontractors are alerted to user need-changes and requirements well in advance of a product's development phase. It is a key element of the target selection process. Early identification gives the prudent industrial organization the lead time it needs to apply seed money (either the government's or its own) in order to begin the early phases of new product development.

SELECTING TARGETS

As we approach the target selection task, it would be wise to review several concepts regarding high technology marketing activities and practices which appear to be rather common among successful companies. We have all heard and read about the practice of protecting a company's core business. This is the business base that, year after year, provides essential operations to keep the company alive and well. Let's define what we mean by core business and how this concept of protecting that core is so important. Then, we can address the overall subject of target selection based on core business, because, for a company to show growth, it must expand outward from this core.

For our purposes here, core business is defined as that business mix you enjoy at any given point in time. It is that program activity that produces the company's net income. The programs are the cash cows, the emerging programs that will sustain the business in the future, and the research and development activities that contribute to the maintenance of the cash cows and the successful transition of emerging programs into viable net income producers. Core business is the business base. It is this base that the target selection process uses as its reference. An inherent part of the business base is the technology base.

The concept of core business and the expansion of that business into various increasing levels of activity is depicted in Figure 9-3. This is intended to show that your first level of expansion must be closely associated with core business; thus, the risks inherent in the movement are minimal because they are real extensions of core business into new opportunities that are directly related. The farther one moves away from core business (toward the outer rings), the greater the risk becomes. Also, the

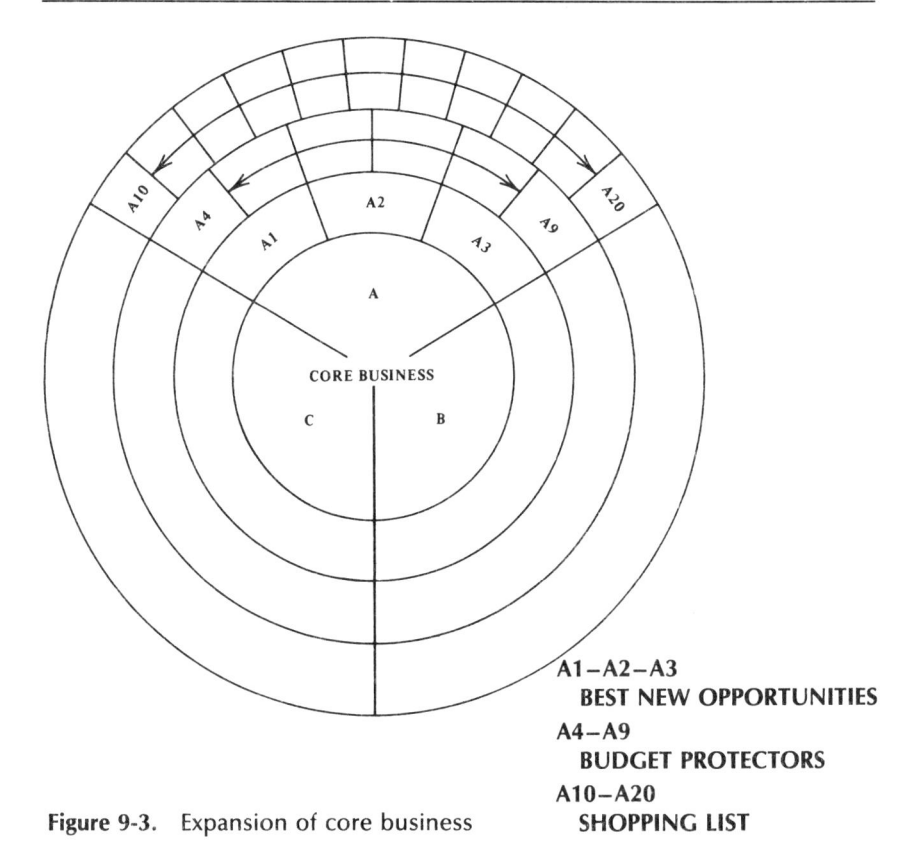

Figure 9-3. Expansion of core business

A1–A2–A3
BEST NEW OPPORTUNITIES
A4–A9
BUDGET PROTECTORS
A10–A20
SHOPPING LIST

farther away you move from core business, the greater the cost of entering that new business segment.

In Figure 9-3, I have shown a company's core business divided into three product line sectors (A, B, and C). I have expanded sector A to show that the marketing group for our representative company selected three primary new business opportunities as *must win*; these opportunities became line items in the company budgeting process. They are the new business program opportunities having the most direct relationship to core business. That is, the technology base, company, customer, and competitor evaluations we discussed earlier show that programs A1, A2, and A3 are the best selections out of the 20 program opportunities evaluated (A1–A20). Programs A1–A3 are programs the company *can win*. Programs A4–A9 are called budget protectors because they do not fall into a can-win category, but instead will require substantially more development (company assets, time, marketing effort, technology advancement) to justify a budget line identification.

Programs (A10–A20) are regarded as programs of interest to our representative company. These are programs usually carried on a marketer's shopping list. They have no leverage in the assignment of company assets for the pursuit of new business.

Product line segments B and C should be treated in a similar manner. For example, product line B may have only one budgeted program, two or three programs in the budget protector ring of programs, and none at all in the outer ring. Product line C may have a totally different mixture than either product lines A or B.

It is from this layout (or any other layout that serves your evaluation process) that you can quickly visualize the key programs out of your target selection process. Programs in the inner ring—the budgeted programs—reached that lofty pinnacle because they relate directly to your core business and are programs you can win. These programs also have first draw on company assets related to new business capture.

A serious mistake made by many marketing managers in selecting their targets for future years is that they will often concentrate on one or two major procurements to the detriment of the other segments of their businesses. This can become so overwhelming at times that smaller product lines can be destroyed; the feeding and care that is derived through continuous new contracts is not considered during the target selection process. The attention-getters (the major new programs) are put on the front burner and attract most of the company's attention and the greatest percentage of new business acquisition assets. This is a dangerous practice and control must be exercised.

In a truly competitive environment, the prudent manager will keep his options open. It is most likely that programs will not be awarded precisely as he has them forecast. Also, even the best of companies does not win every procurement it pursues. Many managers have what is referred to as a built-in *what if* attitude which seems to control their thinking process. It is this skepticism about the high technology marketplace that will often mean the difference between operating a successful business development group and one that is only marginally successful. One marketing manager described his analysis of the marketplace in very succinct terms: "Death and taxes are assured; winning in the high technology competitive marketplace is the last entry I'd ever make on such a listing."

Figure 9–3 can also be viewed as a mental crutch you can use in evaluating new program opportunities. It is my experience that if you examine your new business opportunities with this or a similar graphic in mind, you will tend to be conservative. And, in high technology business, the conservative manager will more often than not emerge as the successful manager in terms of percent of capture relative to the cost of acquiring new business. Given an unlimited budget, or one where practical constraints are not given, it is possible to capture new business opportunities only distantly related to core business. However, the practice does not contribute to a controlled growth and profitability thrust for your operation.

Costs are accumulated quickly when you extend your reach. You accrue considerable cost just to achieve a level of competitiveness. Conservative high technology marketing managers do not allow their staffs to progress much farther away from core business than the first two rings.

In forecasting, it is also the mark of the conservative manager to select only those new targets from within the first ring for the purpose of establishing a hard budget for the future. The probability of capture is greater when you maintain the direct relationship between new opportunity and the core. Once you move away one more step, the probability of capture is lessened significantly.

The one element not depicted in Figure 9–3 is *time*; or, in practical terms, the schedules for programs A1–A20 are not depicted. Let's review how time can become a major factor in target selection. We can assume that program A2 is placed in the budget line category because of its relationship to core business; however, the program will not reach a procurement status for several years. That doesn't change the diagram. The program is simply budgeted for a later planning year. Its positioning tells you it is considered essential to your company's future. Therefore, since you are given lead time to fully develop your can-win posture, you should set forth the details of a capture plan. This may include additional company-funded research and development, accelerated user and customer liaison, greater involvement of all other company disciplines, acquisition of seed research and development contracts to overcome any technology base shortfall, and further study of competitor positioning. Target selection is not limited to current-year budget line items. In fact, if your inner-circle program identifications do not include these futures, it will have a disastrous impact on the future of your company.

For nonmarketing managers in a company, a long list of new business opportunities confirms that growth projections will probably be achieved. Their posture is that with so many possibilities for award, growth is assured. Unfortunately, they are lulled to deep slumber by a reactive marketing function—reacting to intracompany pressure to use *paper* to prove the projections in some business plan. Any marketer worth his base salary can provide opportunity lists. It is only when you begin to evaluate the opportunity for the development of win strategy that you can find overwhelming reasons for not participating in the program under any circumstances. This is a basic target selection misconception. Simple listings—longer is better—is not a viable alternative.

COMPANY PERFORMANCE AND TARGET SELECTION

As pointed out earlier, one of your most important duties is keeping abreast of overall company performance on all of your current contracts. The customer is fine-tuned to contractor performance levels. When you stub your toe and miss major milestones—both technical performance and cost—the reaction within the marketplace can very well dictate how you will score in any subsequent evaluation.

Delivering on time and within cost is a primary objective. But, unless the machine you deliver to your customer *satisfies his need*, your record of accomplishment is tarnished. It is possible for a customer to develop a specification and contract for its development only to learn when it is delivered and installed that some key needs have been overlooked. If your customer cannot use what you deliver to him—even when he is responsible for having overlooked some critical performance criteria—you will bear the brunt of the criticism for delivering poorly performing equipment. Again, it is vital for you to confirm that user requirements are being fulfilled by the equipment you are developing. This can best be done by maintaining a constant liaison with user personnel. His use situation can change without your contractual requirement ever changing. If you have any doubts that a user need is not being adequately addressed, you should review these doubts and concerns with your customer as soon as possible.

WIN STRATEGY

Win strategy is the next logical step following selection of your basic new business acquisition targets. One point should be clear though; win strategy serves as a criterion in target selection. They are complementary. In fact, I believe that budget programs should not be accepted until win strategy is fully developed. Just saying you can win because of a direct core business relationship is not sufficient. There must be a detailed win strategy and a series of stated activities to implement the strategy.

This is why I believe it is important that you resort to the systems analysis concept of marketing. It forces you to examine the total company–customer–competitor spectrum. It is a discipline which I believe is absolutely essential as you develop the win strategy for selected new business targets. The technique will keep you sharply focused on issues that may very well go unnoticed in the panic to prepare your proposal for submittal or even to overcome a technology base shortfall for a future program.

The previous discussions regarding your technology base, customer, and competition are all intended to prepare you for win strategy development. When examined in the context of an operational system, it should clearly indicate the essential elements of your strategy. I'll use two examples to describe how the process is employed. Both examples will cover programs in Category A of Figure 9-3: (1) a program budgeted for current year capture and (2) a program budgeted for capture at current year plus three.

The first example deals with a program budgeted for capture during the current year. Let's assume the selection and evaluation process moved the opportunity directly to a budget line position. This may or may not be good marketing. It may be the result of recent procurement decisions on the part of the customer or of a probe into a new sector of the marketplace. But, for whatever reason, the opportunity is relatively new and you now should decide how you will proceed to convert it into a new order. We'll further assume that the customer is a government agency, their RFP (Request for Proposal) will be issued within 60 days, the competitor listing is lengthy, and the award is expected within 120 days following submittal of technical and cost proposals.

At this point, you need to reaffirm your bid decision. You should carefully review the total decision process you employed to move the opportunity to budget line position. What have you overlooked? Are you considering all of the company–customer–competitor factors we reviewed earlier?

You decide to submit your proposal. The program has content that would mix well with your current business. The agency is one you have earmarked for exploitation. You organize your proposal effort, release bid and proposal funding, and initiate the preliminary activities leading to receipt of the RFP.

The RFP arrives. There are some surprises in the equipment specification. The direct relationships to your technology base have been diminished. Your technical risk expands. The customer has asked for a fixed-price cost proposal while you had been expecting a cost-reimbursable arrangement. You submit your proposal. Many of your key company management personnel are skeptical. You maintain typical marketing optimism.

You lose. Your price is almost double that of the selected contractor. You ask for and are granted a debriefing by the procurement group. You are advised that the technical evaluation portion of the source selection graded you far below the winner. You failed to describe specifically how you would achieve the three main equipment performance objectives given in the specification and also identified in the listing of evaluation criteria. Your reliability prediction was not believable. Your qualification test program failed to include a maintainability demonstration. Your field support plan was weak. Your proposal was difficult to follow; the evaluators had to search for specific responses to the critical elements of the evaluation criteria.

Obviously, this example is a composite of many experiences described by both customer and company representatives. The intent is to show that the imaginary company and its marketing group were not adequately prepared to enter the competition; their win strategy was poorly conceived. Their first mistake was to allow the program opportunity to reach the budget level. Consequently, both target selection and its associated win strategy were improperly developed. It is these marketing management errors that deplete a company's new business acquisition assets. When you determine you *cannot win*, you should stop the preproposal activity immediately. Besides, when a new program becomes a budget line item, you should have already determined that you *can win*.

Now, let's examine our second example—a program budgeted for current year plus three. Again, this is a composite drawn from hundreds of examples. For those who argue that any high technology program identified *now* that is not expected to reach RFP and award milestones for at least 36 months may be a bad target selection, I suggest you read on. Experienced marketers can describe many examples for you. We are now entering the arena of *business development*; this is where we separate from marketing (and sales).

For this example, we will simply refer to the equipment as a device the customer has identified in early planning documents. The technology required to make the device operational is presently still in the laboratory; practical application is on the horizon. The device is essential; it will fill a specific and critical user need. Funding should be no problem.

In evaluating the opportunity based on this rather preliminary overview, the company in our example and its business development group identified several key program characteristics:

Not a severe departure from core business.

A reasonably good relationship to technology base.

Earlier research and development activity funded by the company had addressed similar requirements.

A recent new hire from a competing company had been involved in a similar technology application, and fully understood the user requirement and what technology would be required to adequately fulfill what the customer's technologists had already identified in their early planning documents.

A significant production follow-on potential appeared likely, assuming that the advanced development and engineering development phases could be successfully executed.

The capture of the early phases of the program would expand the technology base in other directions critical to other target opportunities the company was considering.

The company-funded research and development program would require only minor restructuring to better address program requirements.

The customer had sufficient early funding and planned to release several study and research and development contracts to more fully address the requirements.

Only two other competitors had been identified as showing any interest in the program, or even having the basic and essential capabilities to address the requirements.

The customer agency appeared on the company's inactive list of customers. All prior contract work for the customer had been satisfactory from both the customer's and company's viewpoint.

Based on the facts listed, the program became a budgeted target. Win strategy meetings were convened. It was determined that early and essential milestones should include capture of one or more of the early study contracts to be awarded by the customer, and a broadening of the company-funded research and development activity to embrace certain critical technology base issues. It was also determined that an energetic program should be implemented to contact all key customer agency and user personnel to fully demonstrate company interest, capability, and investment. Also, the levels of contact would tend to flesh out the general specification requirement for the company's technical staff.

Here we see a critical business development decision based on the company's viewpoint that the RFP is already *on the street*. The program would now be given high priority and would receive company support (research and development and bid and proposal funds) to develop the opportunity and assure its capture.

During the next two years, the company in our example and one of its competitors received substantial customer funding to address key technology issues and demonstrate the feasibility of moving to full-scale development. Additionally, the company in this example heavily funded its own research and development program to supplement customer funding. Early models of the device were constructed and tested; customer technical personnel were invited to attend demonstrations. Technical papers were prepared to describe the practical application of the new technology. Hints were made in the company's advertising program that newer technology was on the very edge of reaching a practical and essential application. The elements of a very carefully constructed win strategy were implemented. Milestones depicted in the win strategy document were achieved essentially on schedule.

The company in our example was awarded the engineering development phase of the program. The customer elected to bypass the advanced development phase;* they believed (and so

---

* The terms *advanced development* and *engineering development* are extracted out of the life cycle of weapon systems, and are employed rather generally at times by many uninformed writers and others. Department of Defense Instructions divide major programs into five phases, with a key decision required preceding authorization to move to the next phase. The Instructions identify the phases as: (1) Advanced Technology Development (or advanced development, as used in this text); (2) Conceptual Phase; (3) Demonstration and Validation Phase; (4) Full-Scale Engineering Development Phase (or

did our example company) that the technology advancements had matured rapidly through company and customer seed money. Only one other competitor responded to the customer's RFP. The company in our example was selected on the basis of a superior technical and management proposal. The cost packages were essentially equivalent. While no specific measurement standards had been developed, our example company had continued its energetic marketing communications program throughout the proposal and evaluation phases. They were unable to determine if members of the customer's source selection team were in any way influenced by the information released by the company or by the company's direct and timely contacts with all key customer personnel. A later analysis clearly showed that the competitor's marketing communications program never got off the ground. In reviewing the win strategy document for our example company, it was determined that each task and each milestone for that task had been accomplished.

Elements of their win strategy included an extensive risk assessment study (cost, schedule, and technical performance), and their proposal addressed these factors with a detailed program describing exactly how each would be dealt with. The technical proposal addressed every element of the evaluation criteria established by the customer. This was given considerable emphasis in the win strategy. In fact, the technical and management proposal outlines were released nearly six months before the RFP was released by the customer. The proposal team had asked for and received a sufficient amount of lead time to accomplish their task.

While the preceding information regarding our two examples

---

engineering development, as used in this text); and (5) Production and Deployment Phase. Advanced development is considered that segment or phase of a program that will facilitate the transfer of a technology or technologies from a purely research and exploratory development stage into the systems development activity. Normally, in the procurement of large weapon systems, the advanced development phase is intended to provide a range of candidate concepts for further development—and thus for the solution of the problem expressed by the need. Engineering development is considered that segment or phase of a program where the equipment (the system) is designed, fabricated (usually in an engineering rather than a production environment), and then tested to the full range of the applicable performance, environmental, and other specifications.

has been presented in rather simple form and derived from hundreds of real life situations, you can readily comprehend the relationship of target selection and win strategy. There are many who believe that targets should not become budget line items until the win strategy clearly confirms a *can-win* environment. When you pursue your win strategy to the fullest extent, it will more or less dictate which targets should be elevated to the budget line position.

Win strategy—the company posture, attitude, and plan—associated with new business acquisition is generally regarded as an essential part of the marketing process. Companies and their employees at almost all levels of management discuss the concept openly. But, win strategy for any specific procurement is closely guarded; it is often not disclosed except to key management personnel within a company. This is a good example of *executive cranial vapor lock*.

Every competitive procurement you address (and even your sole-source business) should have a win strategy and a detailed plan (activities) to implement the strategy. Everyone in your company associated with the new business acquisition process should be a contributor to this strategic plan and should have specific assignments in any detailed plan for implementation. Winning a competitive procurement is not a marketing job alone. It is a company-wide dedication to the total process involved in the objectives of growth and profitability.

CHAPTER TEN

FORECASTING

Forecasting business levels from the government or other high technology marketplace is much more difficult than for almost every other segment.

One of the primary tasks for the marketing manager and his marketing staff is to predict—with high levels of accuracy—the timing and content of new orders. Once programs have been selected for capture and become part of the future business base (the budget) of the company, all eyes focus on the marketing manager. He alone is expected to meet every objective. In most instances, pragmatic high technology marketers will consequently lean heavily toward a conservative forecast. They are motivated by two good reasons: (1) the customer does not generally meet program release milestones and (2) all other company functions and activities are programmed around the planned release of new programs.

We will review several selected disciplines utilized to achieve improved short- and long-term forecast accuracy. How these *windows* are set and the criteria for setting their position in time will often mean the difference between a successful fiscal year and one filled with uncertainty and mediocre results. The high technology marketplace contains some very fragile elements and must be manipulated carefully to avoid serious business disruptions.

FORECASTING FOR NEAR-TERM PROGRAMS

Good forecasting can be accomplished only if you consider the fragility of the marketplace and the fact that you control very few of the timing factors associated with your selected programs. Forecasting is inexact. Consequently, it is essential to include contingency planning—the alternate actions one will pursue when the forecast line falters, which it will.

One form of forecasting leads you to become so conservative that regardless of the outcome, results are well within the boundaries that have been set. This practice is intolerable in a business enterprise that also seeks profitability and growth. The challenge to do better no longer exists. Objectives are easily achieved, so managers and their groups become lazy and ineffective.

The objective, then, is to fully understand how your customer community operates—how their procurement system works and the timing *they* assign to the elements of the system. Without this essential data base your forecasts become *guesstimates*, and in high technology business (and perhaps most other businesses as well) you create turmoil throughout the total company structure.

The basic question is how can you improve forecast accuracy? How can you eliminate the *guess* element? Is it possible to raise the level of accuracy—to project within some acceptable tolerance when some future milestone will be achieved? We will discuss some very basic elements of the concepts for accurate forecasting in the high technology marketplace. You won't become so skilled that you can predict the future, but you can improve forecast accuracy.

The government's procurement cycles are the most difficult to comprehend and integrate into any forecasting discipline for your company. Whether you satisfy government needs as a prime contractor or subcontractor, the problem of forecasting the procurement cycle with any degree of accuracy is severe and needs to be fully addressed.

Except for standard commodity procurements, government agencies seldom meet their schedules (at least the schedule they establish at the outset). The larger and more complex the procurement package, the greater the slippage in schedule. The procurement offices will almost never beat a planned schedule. Why is this true? The answer is obvious when you examine the flow chart for a major procurement cycle—a chart which includes all of the paper generation and approval cycles, the program funding line and how funds are approved for application to a specific program, the solicitation cycles, source selection process, preaward activities, and finally, an award of a contract.

Earlier, we discussed the selection of new business targets that may not reach a contract milestone for several years. During that lengthy time period, a number of major obstacles are encountered—all of which impact on the contract milestone. The threat can change. (The word threat is used here again in the context of threat to our national security.) The progress of technology will modify long-range procurement plans. The mood of the Congress with respect to government spending levels is unpredictable. National and worldwide economic conditions as well as national politics and foreign policy, play a significant role. Later, we will examine how these long-range programs should be handled—not from the viewpoint of setting precise event milestones, but rather through a series of event windows.

Many business development groups employ major program event flow-charting (or some similar method) to depict the event sequence and timing for their own use and for the use of all other functions within the company. The objective is simply to identify and interconnect all of the key events, from the initial identification of the new business opportunity to the projected date of final source selection and award of the contract. Following the development of the event sequence, the next step is to go back to each event and forecast the elapsed time associated with it—that is, the time it will require to complete the event once it is initiated.

The accumulation of event cycle information for any given procurement is not as difficult as it may appear. Experienced marketers already have a reasonably good comprehension of the steps involved in the high technology program acquisition process for the government and its prime contractors. When you have what you believe to be a true representation of the event cycle, you should submit your flow chart (or other diagram) for review to the specialists within the government agency or the prime contractor facility. Most often, they are willing to examine the chart and make corrections and additions. Some agencies have already developed such charts for their own use and will make them available to you. For the beginner, the government publishes a vast number of materials dealing with the acquisition and budgeting process.* A careful study will soon reveal most of the major milestones and events.

The next step is perhaps the most vital one. Adding timing information to your chart will form the basis for the forecasting procedure for your program. Since you cannot precisely predict event (milestone) occurrences for the future, an acceptable and useful alternative is to apply a meaningful tolerance to all of your selected dates. This is the method employed for project charting under PERT (Program Evaluation Research Technique), or any of the derivatives of this process. The technique requires the application of *optimistic* and *pessimistic* schedule predictions for assignment on either side of your *most likely* prediction.

For example, you may forecast that an event will be completed in June. But, because of program factors, you believe the event could, under the best possible conditions, occur as early as mid April. Conversely, you are also aware of the program conditions that could delay the completion of the event until mid July, or even later. When these schedule windows are applied to all events in a flow chart, you can begin to visualize the wide fluctuations of the schedule for the final event—the contract award. If you have any doubt regarding the width of these event windows,

---

*Of the hundreds of documents, pamphlets, directives, and other publications available, I recommend that you obtain two excellent ones for your marketing department library. The *Department of the Navy RDT&E Management Guide* (NAVSO P-2457) is available through the Superintendent of Documents, U.S. Government Printing Office, Washington, D.C., 20402. Ask for publication Stock Number 008-040-00087-7. The other publication is available through the Electronic Industries Association, 2001 Eye Street, N.W., Washington, D.C. 20006. This document, *Defense Systems Acquisition and Budget Cycles Handbook*, is an excellent and well-organized collection of significant Department of Defense documents dealing with the acquisition and budgeting process.

you should discuss them with your customer specialists. They are usually unwilling to take the time to develop the schedule information for you, but most are willing to review what you have accomplished and offer suggestions for improvement.

When you have completed the chart timing, you can accumulate schedule information using your own best judgment. The total process will provide a better understanding of the program acquisition system employed by your customer, an opportunity to monitor progress of the procurement cycle, and will improve your forecasts for use within your company.

It may not be necessary to apply this technique to all of the new business opportunities on your budget list. Often, repetitive orders for the same product from the same agency can be forecast with reasonably good accuracy. However, I strongly suggest you apply the technique to your most complex targets to gain essential knowledge of your customer's event cycles and to better establish key schedule milestones for the benefit of all other functions within your company. It is your order forecasts which form the baseline for their planning and scheduling activities.

FORECASTING FOR LONG-RANGE PROGRAMS

For your long-range new business development activities, it is often better to establish rather broad windows for key events rather than attempting to complete the total acquisition flow chart with the three time estimates of our previous example. The portrayal of event windows, as depicted in Figure 10–1, serves many useful purposes. In our example, you can enter event identifications within the rectangular boxes provided. At a glance, you can then see the general total program progression from inception to contract award. This method of program milestone projection provides other planning groups within the company with a broad picture of the program cycles. They can begin to drop in their own planning milestones, and thus also contribute to the overall company long-range business planning activity.

Planning charts of this type are of particular value to the research and development planning function. They visually show

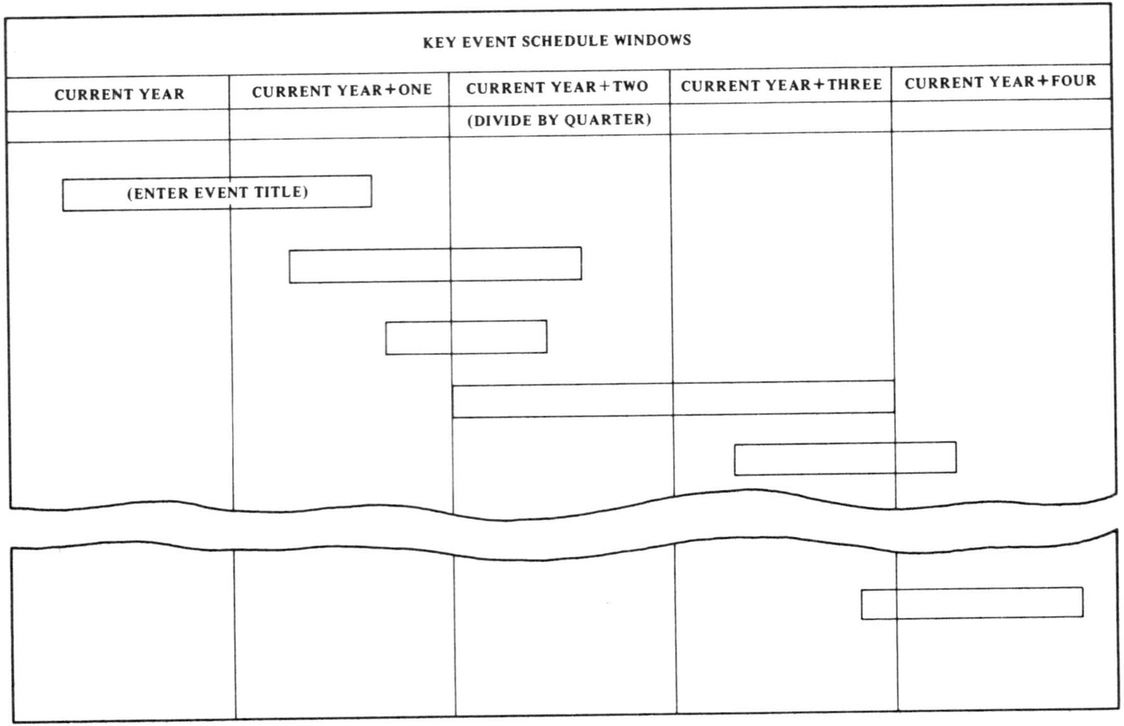

Figure 10–1. Key event schedule windows

when new technology must be available for application and establish key milestones related to specific research projects. These charts are also helpful in planning technology acquisitions, licensing for the application of technology held by others, facility and manpower planning, and capital investment planning.

I believe it is wasteful to attempt to tie specific calendar dates (within a month or so) to long-range programs in the high technology business environment. Program targets that are expected to reach the contract award milestone in three, five, or more years are subject to every possible variation. In fact, some targets will even completely disappear, or at least they will be combined with similar and related programs.

THE FLEXIBILITY OF FORECASTS

A forecast isn't worth the space it occupies if there is no plan for achievement. You must continually question the soundness of your plan, test the plan against progress at selected checkpoints, and be prepared to modify the plan, *and the forecast*, if conditions develop which call for a modification. Assessment (or evaluation) of progress against plan is essential. To move ahead blindly—because that's what the plan requires—is the surest way I know for failure. Furthermore, adjust your forecasts whenever new information becomes available since other company functions also need lead time to perform their assigned tasks. The best possible accuracy of your forecasts is essential, so you must be alert for rapid change. All functions within your company should also be prepared for change and able to make it a part of their planning system.

CHAPTER ELEVEN

PROPOSALS AND PRESENTATIONS

Proposals in the high technology marketing environment take many forms. These may include solicited and unsolicited technical and cost proposals, and any other document intended to influence an award or in any way modify, restrict, cancel, or supplement one in process. It has been said that any document transferred from a company to its customer is a proposal since in some way all documents reflect the company's ability to perform. If this is true, then the marketing manager must play a significant role in the preparation, review, and approval cycles associated with these documents. The company–customer relationship is of great concern to the marketing manager. Consequently, any proposal should be given the attention it deserves.

We will review basic proposal types (from major technical and cost submittals to simple letter-type proposals) and describe the critical role of the marketing manager and his staff. While they may not generate much of the content, they are certainly responsible for content and format that will be responsive to the customer's request and need. In a highly competitive environment, the proposal (in any form) may be the only evaluated formal link between the buyer and the seller. Awards are won and lost simply on the basis of proposal content. On major programs, competitors may each spend millions of dollars of their proposal assets to prepare responses to a solicitation. Control of these company assets is essential.

The proposal is the marketing manager's product. All of the control aspects of production, quality, and user need we see for hardware items can be equally applied to proposals. My objective is not to describe in detail how to write the proposal, but rather to provide some tried and proven methods for managing the effort to obtain best possible cost and technical presentations.

IMPLEMENTATION OF PROPOSAL TECHNIQUES

Much has been written and said about how to prepare winning technical and management proposals. Adams, Close, Silver and others have for years given exceptionally fine seminars on the subject. Hundreds of technical and management proposals have been improved and have won vital and important contracts for companies thanks to these teachers. However, the bottom line is implementation of their teachings—actually accomplishing in your proposals what they tell you should work best for you.

My own analysis shows that one of the most significant reasons technical and management proposals don't work for companies is marketing management's inattention to the detailed proposal preparation process. A great percentage of the blame for a losing proposal can be traced to top level management personnel in the company who failed to provide essential leadership during the preproposal and proposal phases. Business development (marketing) managers, then, must take the initiative to develop the right kinds of proposal awareness and leadership throughout the company. For some it will be difficult. For others it will be a restatement of lessons already learned, but with greater emphasis on implementation.

Marketing managers traditionally look to their proposal manager to lead the effort. He is usually an unwilling leader plucked out of the company organization. Often, he is pulled away from some equally important duties only because he is believed to be the best employee available at the time the proposal effort is staffed. Suddenly, he becomes the single most important element in the development of the winning proposal. It is, in my view, one of the most serious errors in proposal activity. However, it is common practice for many companies.

By arguing successfully, some new business development managers have obtained specialists for their companies whose only responsibility is the management of proposal activities. These people are skilled, know how to respond to a request for a proposal, and are intimately acquainted with the company's technology base and current programs. Because they are solely responsible for proposal preparation, they already have first-hand knowledge of what the customer wants to see in a proposal. They devote their energies to doing exactly what is required of them. They maintain direct contact with the marketing and technical specialists who have been working with the customer, read and study marketing plans, examine competitor assessments, and assist in the development of a win strategy for the proposal effort. They are also aware of other related activities within the company and can select what they need to develop a winning proposal.

An alternative to the single-point specialist for the preparation of all proposals is to identify individuals from within the company who will serve as proposal managers. These assignments should be made as soon as programs reach budget level. That is, once a program has been identified for capture, someone in the organization should be identified as the proposal manager. Early in the procurement phase, he can continue with his regular assignment while devoting only sufficient time to the new program in order to begin to develop the dossier he will need later. In this manner, he is given lead time.

This is a good practice for many smaller companies who cannot afford to maintain full-time proposal specialists. It is certainly much better than suddenly thrusting a major proposal effort on an individual who is already working to full capacity on his current assignment. One or the other will suffer; more often than not, it usually yields a mediocre proposal.

PROPOSAL RANKING

I have concluded, after many years of participation, that cost–management–technical proposals are most often arranged in this specific order in the determination of whether you win or lose. This arrangement is not intended to show that technical proposals are not important—they are. They form (for almost every procurement) the baseline for the customer's evaluation of your technical competency. However, in today's highly competitive business environment, most of the competitors for any major procurement you may wish to select are technically competent. Otherwise, they wouldn't appear on the list of competitors.

Next in order of importance is the management proposal. Generally, this is the portion of the proposal package that describes how you will perform the managerial tasks associated with the program or project. Here, you are selling the total company. Procurement officers want to know that if you are awarded the contract, you can reasonably be expected to manage the tasks outlined in the contract. Many companies in the high technology business area prepare outstanding management-level proposals. They address every single issue of importance to evaluators. They have learned (often through the experience of failing the source selection process) that the management plan for any given procurement is vital—for the company *and* for the customer faced with the selection decision.

And, finally, cost. Most procurement decisions are based on the cost proposal. Are you competitive? Are you the low bidder? Are your costs reasonable? Are your cost packages based on poor estimates or are they supported by factual estimating data? Are the risks properly assessed and priced? Are you really prepared for a preaward audit? Are your costs substantiated by similar program experience? Credibility is essential!

MANAGING THE SMALL PROPOSAL

Smaller proposal efforts, requiring minimum management and technical input and simple cost/price structuring, do not normally require a major involvement of other company functions, but a discipline for their preparation should be developed and implemented. Every proposal should follow some structured plan for preparation, approval and submittal.

One simple solution for you to consider is the restructuring of the disciplines for major proposals. Strike out the nonessential steps. What remains will give you sufficient control over the preparation and approval phases leading to submittal. If the new business opportunity is worthwhile (you shouldn't submit a proposal under other circumstances), then it logically follows that you should have a procedure covering the work flow.

In the high technology business environment, many small proposals (reduced cost and technical complexity) may be submitted to obtain research and development study contracts and seed funding for exploratory and advance development tasks. They are vital to the long-term success of your company. They should be viewed as essential early steps in the total acquisition process for some identified major target program. In most instances, the structured routine described here, while more rigid than that generally in practice, should increase your capture percentage.

PROPOSAL CONTENT

The following is an excerpt from a government procurement package regarding the development of your management and technical proposals:

> Unnecessarily elaborate brochures or other presentations beyond that sufficient to present a complete and effective proposal or quotation are not desired and may be construed as an indication of the offeror's or quoter's lack of cost consciousness. Elaborate art work, expensive paper and bindings and expensive visual and other presentation aids are neither necessary nor wanted.

A word of caution is necessary here. Don't be misled into believing that words like *elaborate* and *expensive* can be ignored. Cost control is important. However, the skilled graphic arts and proposal preparation group in your company can turn the proverbial sow's ear into a silk purse by utilizing style and an effective method of preparation. Proposal content and the method of presentation are usually not enhanced by better paper and elaborate artwork. A little pizzazz is necessary; but don't let the pizzazz overshadow the real intent.

Evaluators are uncanny in their ability to see through the gloss to uncover the incompleteness and ineptitude buried beneath. Make it easy for them to find the answers to their questions. Bring your strong points to the surface where the evaluators will see them; guide them to your specifics in response to their evaluation criteria. This is no place to play hide-and-seek; evaluators are inclined to assign low scores to proposals having this characteristic. Again, a little pizzazz is good, but don't let it substitute for the real message. The real message must be clearly visible.

THE COST PROPOSAL

Up to this point, we have reviewed generally some of the best disciplines relative to the management and technical proposal. They are important. However, more contracts are won on the basis of price than they are on the relative merits of management and technical proposal content. True, you must strive for the competitive window with your proposals; but if you do not achieve the competitive window on the price, you will not become a finalist in the process leading to contract award.

There are many facets to the process of developing your best cost structure for a target program. To begin, you should have a reasonably good estimate of *should cost*—an often elusive number based more on experience than on actual estimates prepared by various operations in your company. When you selected the target and developed your win strategy, cost should have been one of the key items you considered.

A *can-win* attitude reflects the favorable impressions of management and technical proposal posture. It also reflects a general company attitude that your cost proposal will fall within the competitive window. First, you know approximately how much the customer is willing to spend. Program budgets for direct government procurements are not all that difficult to ascertain.* Then, you have (or should have) a reasonably good understanding of the many cost drivers inherent in the program, including knowledge of the risk areas within you own design approach. Your competitor files can (or should) provide you with a reasonably good estimate of how they may structure their cost packages and develop their final price. Given these various inputs, you can determine, in logical sequence, how to structure your cost package and develop your final submitted price.

*Government program budget data aquisition can be a complex task for someone who has never tackled the federal bureaucracy in Washington. The first problem you will encounter is the relative inaccessibility of program offices, personnel, and documents containing estimates for your programs of interest. It becomes a much more difficult task when you seek budget estimates for low priority line items buried within a total departmental budget.

Experienced marketers and companies both large and small with broad experience in government contracting have developed *points of contact* within the sectors of their marketing thrust. This is not intended to suggest that government proprietary budget data (estimated costs) for products and services are distributed to favored competitors.

It is important to understand that when government agencies solicit budgetary or not-to-exceed preliminary job estimates from industry (and not all potential competitors participate), they are setting the high/low limits for the budgeting process.

Government program office personnel assigned to your long-range programs of interest have a responsibility to assist in establishing realistic job cost predictions for purposes of accumulating data for the annual budget submittal—and for the years to follow over which the program will require funding. Becoming involved in these early planning activities is vitally important if you intend to continue to pursue these programs; it gives you the trail for the budget process. (For the newcomer, I strongly suggest you review the federal budget process described in the documents recommended in Chapter 10.)

As already stated, tracking your programs of interest can be a very complex task. However, let me clear the way somewhat by reminding you that you can probably ignore over 95% of the federal budget and all the related line items; that portion has absolutely nothing to do with your area of interest.

It is imperative that you learn to speak the language of the customer. Learn his system. Thereafter, I suggest you employ the most basic marketing approach. You ask—and you don't stop asking until you have achieved your objective! Earlier, we reviewed the requirement for broad company involvement in the marketing task. Every customer contact made by anyone in your company should include a requirement to ask a series of questions related to program details, the competition, user problems, need—and funding availability. Ask! Whatever I've said here applies equally well within industry. The techniques are the same.

Your targeted customer may be unwilling to disclose his budget for the product or service you expect to sell to him. However, he will often provide a *window* for you to shoot at in your pricing submittal.

The low bidder wins. How often have you had that fact confirmed? (True, there are some outstanding examples where major procurements have not been awarded to the low bidder.) You can be the low bidder for most of your selected new business targets if you follow the target selection and win strategy development process reviewed earlier. That process should define—with fairly accurate results—your cost posture for selected programs. Consequently, you are forced to examine your competitive position in many different ways. When your selected targets reach budget program level, you have already determined that you can be competitive. All that usually remains is the completion of the estimating process. There should be no major surprises.

With respect to the generation of your final cost proposal, I don't believe I can provide any additional guidelines beyond what is already common knowledge among the most successful business development managers. Companies have their own reasonably well-developed procedures for the process of accumulating cost estimates. My only suggestion is that for whatever system you employ, there must be a method for accomplishing the *internal review process.*

The cost packages generated by the various company organizations must reflect the best possible estimating procedures. Otherwise, setting the selling price—the final management decision process—might just as well be accomplished by a throw of darts at some imaginary pricing matrix. In the final analysis, company management must determine the desired margins and price each job accordingly. Poorly constructed cost estimates will destroy any margin-setting decision process. Moreover, the objectives of profitability and growth may never be fulfilled.

One further factor of cost estimating needs to be addressed here. More and more, we are being confronted with the elements of DTC (Design to Cost) and DTUPC (Design to Unit Production Cost) in the high technology business environment. This is simply a procedure for establishing early cost objectives. These cost projections are extremely important, and can therefore become a heavy burden for the company if they are not carefully constructed.

Many companies are establishing separate cost estimating groups to address these issues and to develop their cost projections under a very disciplined and logical methodology. Most experts agree that your success in meeting these long-range cost goals is tied directly to an informed and participative design and development organization. When that organization understands the competitive nature of the marketplace and the real purpose for establishing meaningful and realistic cost objectives, the balance of the DTC and DTUPC disciplines seems to be achieved rapidly. Then, the only essential task which remains is the review process for the purpose of measuring progress and assigning corrective action where it may be required.

SPECIAL PROPOSAL CONSIDERATIONS

The single most important objective of the proposal is to convince the customer that your company is the best among many to perform the task for them. But, remember that the hyperbole of your proposal is usually consumed at the lower echelons of the evaluation task group. As the source selection decision process moves upward, the written proposal factors become less tangible. Then, you are evaluated on the basis of opinion, judgment, and the perception they have of your company. Your primary task, then, is to shape their perception of your company—to attack their senses at a strategic period of time. This is the final proposal process. It is simply the manipulation of the thought processes of the people charged with the responsibility of making the decision for or against your company for the award. All else being essentially equal, the award decision strongly favors the company with the most efficient marketing communications program. Later, I will review these techniques and the vital role of the marketing manager.

Two remaining points need to be raised here, if only to serve as a reminder for the marketing manager. Win or lose, you should schedule a debriefing with the procurement agency or group. Most marketing managers will schedule a debriefing when they lose; few remember to schedule such a meeting when they win. The purpose should be obvious. You are preparing for your very next proposal submittal, and anything you can learn from prior success or failure needs to be added to your overall proposal preparation process. The other point relates to post-award marketing.

Marketing managers and their staffs have a tendency to walk away from both a winner and a loser. Instead, they look forward to the next opportunity on their list of targets. However, the current contract with that new (or continuing) customer may have additional opportunities which require further development. For example, spares, field support, training, test equipment, spin-off products, product improvement, and technology insertion potential all become available to an awardee. The marketing manager's role in the acquisition of new business must include these opportunities. At least, he should confirm that the company is properly structured to acquire much of this after-market potential.

Many marketing managers make the mistake of viewing proposals as the formal documents submitted in response to the request of a customer for a quotation. Every bit of correspondence, every presentation, every preliminary qualification document, every meeting, and every telephone conversation between you and the customer are part of the proposal process. The final proposal submittal should be nothing more than the formal response to his formal request. It should confirm, in the form of a document, what he should already know about you. The proposal effort starts at the moment you first discover the new business opportunity. If you maintain that perspective, your formal proposals will improve and your capture percentage will increase dramatically.

This is the *balanced attack*! Winners in the sports arena possess a balanced attack and the commander on the battlefield requires a balanced attack if his campaign is to be successful. When you win a targeted procurement it is not through the intervention of Lady Luck. It is the culmination of fitting together many pieces of the win puzzle.

CHAPTER TWELVE

RISK MANAGEMENT

The high technology marketing manager (as all other managers) is required to evaluate business risk prior to the decision process. What may be misunderstood by high technology managers is that the very nature of their business may generate far greater risk for their company than that encountered in other business enterprises. Thus, the identification and subsequent management of risk is elevated to a higher level of importance.

Research, development, and production for high technology products contain severe risk elements. How they are identified is as important as when they are identified. Carrying partially reduced risk forward into each subsequent phase of a program is not unusual. What is essential though is that the risk be identified and tasks assigned to reduce the elements of the risk. The magnitude of the risk will often dictate the *stop/go* decision process. Having a reduction plan available for the decision process is an essential part of the manager's job.

In this chapter, we will provide definitions for the manager and a framework for risk identification, evaluation (assessment), and management.

RISK DEFINITIONS

Simply defined, risk is exposure to the chance of loss. Loss is measured in terms of profit dollars, as well as company prestige in the marketplace.

Most successful companies have a discipline for risk analysis. They strive to identify business risk; they also try to build in a certain amount of protection against its negative impact. But, risk analysis may be left to the designer of the new product. Consequently, when risk analysis is not allowed to surface for comprehensive and effective management control and corrective action, the entire structure of the business is weakened.

Therefore, in any discussion of risk, I much prefer to employ *risk management* as a more descriptive term. Inherent in this term are several essential and effective disciplines to reduce (but never completely eliminate) the chance of loss.

Risk management is an essential part of high technology business. The application of new technology to new products constitutes a measurable risk. There is considerable risk in setting target dates for the emergence of new technology from the laboratory. There is risk in setting target dates for the practical application of new technology. And, there is risk in not having the capability to set reasonable target dates. This in itself is one of the key factors in the competitive environment of high technology application. Therefore, one can conclude that to become more competitive, a company should establish a formal risk management discipline. The chance of loss includes the loss of a competitive procurement.

Risk identification is the essential baseline requirement for a successful risk management discipline within your company. It extends from the designers in your laboratory to the material acquisition and production specialists. Since risk is exposure to the chance of loss, it logically follows that exposure can occur at any point in the organization. Traditionally, risk in the high technology business environment is associated with *the premature application of new technology or its misapplication to an operational problem.* Loss is seen in performance, which impacts on schedule, which impacts on cost. This interrelationship is standard. The basic and immediate problem though is the establishment of a company-wide discipline that will allow risks to be identified. For without identification, there is no risk management program.

RISK IDENTIFICATION

One of the most important aspects of target selection and the development of win strategy is the identification of risk. Marketing managers can lay out the broadest possible array of new business opportunities, but unless risks are identified for each of the targets, it becomes very difficult to continue with the selection process.

Risk, once identified, must be converted to a recognizable quantity. The decision to apply resources to reduce or eliminate risk is dependent upon the magnitude of the risk and the assets which must be applied. Time is obviously also a factor. For some procurements, you may determine that your lead time is insufficient and any amount of asset application would not position you favorably in the competitive environment.

Unfortunately, for many marketing managers, the whole process of risk management is overlooked until the formal RFP or RFQ (Request for Proposal/Quotation) is released. Then, the obvious occurs. Whatever risk is inherent in that procurement is reflected in higher cost estimates (to provide some protection against loss) or is not identified and thus may result in severe losses following the program award. However, marketing managers are protected somewhat by the procurement system. Evaluators will often eliminate high risk competitors before they shoot themselves in the foot with a formal contract agreement.

MANAGING RISK

Much is written about the successful company manager. One of his attributes is that he is willing to take a risk. However, I have yet to meet the chief executive officer who is willing to take a risk when he doesn't also have a program for reducing it. Risk-taking is part of good management, but so is good visibility.

Risk management programs will provide timely risk identification and evaluation. And, when risks are identified (regardless of magnitude) at the time of the original target selection process, you provide lead time for the development of a sound, practical risk reduction program.

In the high technology business environment, it is not unusual to carry risk into a program after the contract has been signed. You may or may not have full protection against loss. Cost reimbursable contract arrangements provide some protection. However, you can control the magnitude of your loss through your risk reduction program.

Marketing managers are forever complaining about the highly competitive nature of the marketplace and the problems within their companies regarding the achievement of truly competitive cost proposals. Quite often, they are not competitive because their cost estimators are adding risk protection labor and material dollars to their estimates. True, these estimators have identified risk; true, they have allocated resources to protect against loss. The problem though is that only they know (hopefully) the magnitude of the risk and whether or not they are adequately protected. By establishing a risk management program, you take the problem away from the estimator and move it into a company-wide system designed to cope with risk conditions. It provides the essential visibility.

Once a new program is under contract, you may encounter new risk that even the most expert analyst would not have predicted. Again, early identification is essential. The best technique you can employ is the program review process. Project leaders for every discipline are required to stand before their peers and appropriate company management personnel and provide their best assessment of current status. If properly conducted, these reviews reveal certain trends that are often the seeds for tomorrow's risks. The best project leaders for your company are those who provide an early warning that risk may be encountered. Recovery programs can thus be implemented immediately to protect the project. Once risk is out of control, it can have a devastating impact across the entire organization.

Marketing managers should remain alert for these new risk encounters. Often, a current project is the foundation project for a broader market thrust. And, since marketers are generally the most optimistic among all company personnel, it is essential that they be given full rein only when it is prudent to do so. When the foundation project encounters difficulty, the marketers should participate in the evaluation process to better construct the marketing activity.

OVERLOOKED RISK

Up to this point, we have discussed obvious risks in the high technology environment. Premature application or misapplication of new technology is the most severe form of risk you will encounter. Other types of risk will also be confronted which often defy identification. However, experienced marketers and other company management personnel have usually encountered such risks before and will look for them in any new procurement.

Long-term new business opportunity targets in the high technology government area may completely disappear. To name but a few of the planning pitfalls, need is never fully confirmed, the requirement is folded into a companion procurement, or funding requests are not approved. At risk then will be your front-end investment in general marketing assets, including the valuable research and development fund. A certain level of early control can be established through the generalization of research and development funding application, whenever possible. The effort undertaken should have two or more identified targets; thus, when a planned procurement disappears, some (if not most) of the investment has direct application elsewhere.

A general risk condition for business is the tendency to slip into a superficial planning mode when business conditions are good and the net income line is fat and comfortable. Experienced managers in high technology industries have witnessed and survived the cyclic nature of their business, and generally incorporate strategies to overcome periods of downturn. Notable among the strategies these managers utilize are the pursuit of the acquisition program (seeking opportunities to bolster the business for the long term) and the establishment of licensing agreements for the exchange of technical expertise. Business cycles often impact on large segments of the industry. Thus, when it is to your advantage to conclude acquisitions, licensing, and technology interchange agreements, you should not hesitate.

It has been said that a willingness to take risk also confirms a willingness to accept a certain amount of failure. Risk management, then, can also be said to be *failure effect management*. The objective is to reduce effect to an absolute minimum. Implicit in a good risk management program are several alternatives to any solution path which is selected. Often though, you will have no need to resort to your backup or alternate solution plan.

High technology marketing managers should insist on a broad and effective risk management discipline within the company. And, the total new business acquisition group within any high technology company should play an active role in risk identification and the evaluation of solution alternatives.

CHAPTER THIRTEEN

CONTRACTS, PRICING, AND NEGOTIATION

The previous chapters have set the stage for this discussion of the types of contracts you will encounter, how pricing strategies are formed and applied, and how you should prepare for the negotiation of the final contract price and terms.

For the intraindustry supplier, government contract types are usually of little interest. However, the partial flow-down of government contract provisions from the prime contractor to subcontractors and finally to lower tier suppliers is common practice. Thus, suppliers at all levels should be familiar with government procurement practices. These practices are the framework around which high technology industry functions. Many suppliers will only see provisions as they are listed in the fine print on the purchase orders they receive from the customer. But, for the majority of the organizations addressed here, government contract types and provisions will accompany most statements of work and specifications received from industry.

Pricing and negotiation practices do not show a wide variance within the high technology business sector. Perhaps this stabilization is a function of the procurement guidelines (the government's acquisition regulation) most companies employ as the primary source of guidance. Also, within industry, customers and suppliers often swap roles and learn from each other. Today's customer can become tomorrow's supplier. In many instances, one can find companies serving both roles at the same time—for example, Company A serving as both a supplier to and a customer for Company B.

In dealing directly with government procurement groups, the subjects of pricing and negotiation follow written as well as unwritten rules. The problem becomes one of knowing your customer and his attitudes. This will enable you to gain the best possible contract for your company and your customer for each procurement.

In my view, the one aspect of contracts and negotiation often overlooked by many managers is the last statement—both you and the customer should be fully satisfied with the final contract. The final contract is an agreement to jointly participate in a program. Each participant should strive for contract and pricing terms that will allow him to achieve his stated objectives. It is a partnership. A program or project should not begin under any other circumstances.

THE GOVERNMENT PROCUREMENT (ACQUISITION) SYSTEM

To be effective, marketers at all levels in a business development (marketing or sales) organization should become familiar with the government's acquisition system and practices. It is an essential job requirement for the government marketer; it is extremely beneficial for the marketer in the lower tier of subcontractor levels.

Generally, marketers do not negotiate contracts for their companies. (There are perhaps exceptions in smaller organizations.)

As a general rule, company staffs include contract specialists who negotiate contract terms with the company's customers. The purpose here is one of defining for the marketer how well he should understand the government's system and how that knowledge is of great value to him.

Contract specialists in most high technology organizations usually do not become involved in the early activities of marketplace studies, company technology base evaluations, and new business target selections. Most program selections (often years before the contract specialists are called in to negotiate) already have a basic framework for acquisition—a preliminary plan for the procurement which the government or a prime contractor expects to follow. The marketer should at least understand this early plan; otherwise, the target selection process may include a program or programs which his company may not wish to pursue. Research and development under a fixed price contract arrangement is a prime example. This will be discussed later.

Marketers (including the marketing manager) can become adequately acquainted with the government's acquisition system by attending seminars devoted to this topic or by simply studying the directives included in the two documents cited in Chapter 10. Seminars related to high technology and the government acquisition system are conducted throughout the year at many locations. Many of them are sponsored by the Technical Marketing Society of America. These seminars are conducted by leading government and industry spokesmen and they review and discuss a broad range of topics—all of which are of interest to the technical marketer and marketing manager.

For specific details, I strongly suggest that you learn how to use the Defense Acquisition Regulation (DAR). The DAR is available from the Government Printing Office or through the Defense Contract Administration Services (DCAA) office for your area. Most high technology organizations doing business with the government already have these on file, or subscribe to Government Contracts Reporter, a voluminous and highly regarded service provided by Commerce Clearing House, Inc., Chicago, Ill.

It should be noted here that in March 1982, President Ronald Reagan signed an executive order to put the administration's stamp of approval on a proposal to consolidate all government procurement under a single umbrella. This is called the Federal Acquisition Regulation (FAR). I doubt that any of the present DAR will be modified—at least to the extent that marketers should be immediately concerned. It is expected that the differences between the DAR, NASA, and other nondefense procurement practices will narrow. Thus, marketers and contract specialists will find that, in the future, the task of understanding and implementing the regulations will become a little less difficult. This is a major project with many unforeseen roadblocks. Some experts remain skeptical that a FAR, which all government agencies can adopt, can be achieved in the near term. But, at least the foundation has been set and significant progress has been achieved. Most companies already subscribe to publications available through the Office of Management and Budget (OMB) and the Office of Federal Procurement Policy (OFPP) in Washington to keep abreast of the proposals and their status.

Another excellent source of contract management information and guidance for practitioners is the National Contract Management Association. Its monthly publication, *Contract Management*, is regarded by many contract and marketing managers as "worth its weight in gold" because in the articles both contractor and government contract management experts provide information and insight not readily available elsewhere. Open discussion of the government acquisition system is an excellent means for gaining a better understanding of a very complex subject.

CONTRACT TYPES

The various contract types have received good general coverage in the literature. For the marketing manager, this level of information may suffice for his day-to-day activities in the marketplace. The basis for all contract type discussions is the Defense Acquisition Regulation. I have never suggested that marketers should become experts in contract law and management; I have never suggested the marketer should study the DAR from cover to cover. There are too many elements involved to necessitate that the marketer or his manager become as fully informed as the government or industry contract management specialist. But, a certain amount of basic information is essential; knowing and understanding various contract types will help guide the marketer through the marketplace.

First, the fixed price contract states that you will perform the described work package regardless of the ultimate total cost to your company—except for any negotiated changes. It is therefore extremely important that your company's pricing for such contracts be based on the best possible job cost accumulation. Contracts of this type should not be accepted unless the job costs can be estimated with a high degree of accuracy. This means that all of the job risks have been considered, all contemplated tasks have been properly estimated, and the impact of projected fluctuations in both labor and material costs have been thoroughly examined.

Here we can see the emergence of the classic example of the utilization of a fixed price contract under some of the worst possible conditions. Both industry and the government procurement system must share the responsibility for allowing the condition to develop.

Since fixed price contracts require you to know—with a great degree of accuracy—what your total job costs are likely to be, a company cannot (or should not) accept work where the risk of overrun (and thus increased cost) cannot be protected. Yes, it is possible to price a risky job in such a manner that all contingencies will be covered. However, you won't win many of these. The classic example is the research and development (R&D) contract.

The R&D contract requires you to perform certain investigative tasks to prove feasibility—to prove whether or not a stated need can be satisfied by applying known and emerging technology. It is risky. Job costs cannot be accurately predicted. Yet, many procurement specialists will hold firm in their belief that the supplier (your company) should be smart enough to predict the outcome and thus include sufficient dollars to pay for all contingencies. As I've stated before, I believe the contract reflects the partnership—the joint endeavor—you and your customer have developed. Procurement specialists must look to their suppliers for an honest appraisal of the cost risk. Suppliers must provide their best assessment of risk and be more than willing to accept fixed price terms and conditions when it is prudent to do so. It simply becomes a matter of casting aside the adversary relationship to jointly develop the best terms and conditions—and contract type.

The management teams for the vast majority of high technology businesses will not allow their organizations to get locked into fixed price contract terms for a development program where high risk is involved. Most companies follow very strict guidelines with respect to the maximum levels of risk they will accept. Thus, an uncompromising procurement office seeking a fixed price contract for a high risk program will (and should) lose a substantial portion of competitor interest. Competitors simply walk away from such risky projects and apply their marketing energies to procurements promising better terms. What is left then is a competitor cluster that does not include the best potential contractors.

Certain modifications of the fixed price contract serve both the contractor and his customer. Economic price adjustment clauses provide some protection against the unpredictable fluctuations in both labor and material costs. Contract prices can be adjusted for either increases or decreases from some previously established level. Often, contracts of this type will contain a ceiling to protect the customer. A redetermination type contract provides for the submittal of a revised cost proposal following the completion of some percentage of the work effort when a more accurate total fixed price can be determined. The final price of the contract is then negotiated, and can be higher or lower than the original submittal.

A fixed price incentive contract contains some of the same rationale as for the reimbursement type. However, the incentive contract is structured around your submittal of a target cost, a target profit, and a formula you and your customer can employ to determine the share of the differences between targeted and negotiated final costs. The formula is structured in such a way that you are rewarded for better cost performance (costs below target cost) and penalized by loss of profit if final job costs exceed the target. The cost sharing formula can have many variations. Incentives can also be established around schedule and technical performance. The varieties and formulas established for incentive contracts are almost unlimited—or at least limited only by the imagination of the procurement office personnel and the company's contract managers.

Cost reimbursement contracts also have several variations. The cost-plus-fixed-fee contract is the most common in this category. It simply means that you and your customer agree on an estimated cost for the job and a fixed fee (profit) you will receive, regardless of whether your actual costs exceed or are less than the estimated costs. The cost-plus-incentive-fee contract features an agreement on target cost, a target fee, and the incentive formula for the determination of the final fee. Such a contract, unlike the provisions of the fixed price incentive type, provides both a minimum and a maximum limit on the adjustment of the fee.

The cost-plus-award-fee contract consists of a base fee, which does not vary with contract performance, and an award fee added to the base fee on the basis of the positive evaluation of contractor performance against any agreed-upon measurement criteria. Quality, schedule, technical performance, and cost effectiveness (including total cost of ownership for the customer) are often cited.

Lastly, the time and materials contract is quite common. You negotiate for the reimbursement of labor hours (burdened plus a profit) plus the actual costs of the materials required for the performance of the described tasks.

One other contract methodology should be discussed under this heading, although most organizations will not be confronted with the issues involved. This is the controversial multiyear contract which usually falls under the fixed price category.

According to its proponents, multiyear contracting is vital in terms of reducing the total cost of acquiring a product or system by allowing a contractor to plan for and commit (to his suppliers and in-house production entity) a total quantity for manufacture over a period of years. Three- and five-year programs are most often cited. The economies should be obvious. Advance procurements of various critical components of a total system (to fill needs for the duration of the contract) yield substantial savings, even for low value items. A single set-up cost and an efficient run-rate over a defined period of time will reduce cost.

At the present time, the controversy revolves around the government's liability regarding start-up costs and work in progress in the event of contract cancellation; whether funds for the multiyear period should be available at the outset or should the contract be funded by annual appropriation approval; and whether the concept should be reserved for major military systems or should it also include the entire range of procurements, down to pencils and paper clips. Also, an existing dollar limitation ($5 million) is considered too restrictive. The programs capable of providing the forecasted cost savings would need to have that limitation raised to $50 or $100 million, or even higher.

Marketing managers should become familiar with the issues involved. A great deal of publicity is being given to the concept, and perhaps the changes many government and industry procurement experts believe are essential may eventually be approved and become general practice in the future. The real issue is our collective ability (the government–industry partnership) to adequately plan for the future—to implement the multiyear contracting choice only when the plan for the future is based on the best possible assessment of need over the long term.

PRICING STRATEGIES AND EFFECTS

An important aspect of contract pricing is seen for that segment of your business where you expect to obtain repetitive orders for the same product. At the outset, you should strive to establish and negotiate the best possible (highest price with best margin) contract terms. A good example is the annual procurement for one of your products which serve a subsystem component requirement in a total system to be procured over a number of years. By negotiating the best possible terms on the original procurement, the customer's buyer–negotiator will then have a basis for subsequent awards to you. Negotiations such as these will usually allow some percentage increases over prior years because of increased material and labor cost—these escalators are often tied to some broadly recognized index of economic conditions. But, how does all of this help your situation?

The key to continued high margin, and thus profitability, is seen in your company's capability to produce the item at a lesser cost. As you progress from year to year, building the same item, your production costs (calculated on the basis of first-year contract dollars) should steadily decrease. You should become more efficient. Your ability to produce the item at a lower cost should steadily increase. Thus, you can see that first-time negotiations, under the conditions described here, can be vitally important.

There is one point that is overlooked by many experienced marketers. There is a strong tendency to submit the lowest credible pricing for a job that will be awarded with cost reimbursable provisions. The price is *cut* to win with the expectation that overrun proposals will be funded later to recover most of the costs incurred—and which were probably identified at the outset.

If you elect to pursue this course of action, you must still thoroughly understand the full impact on your operation when the cost overrun becomes a reality. Contractors often make the mistake of relaxing cost–schedule–performance control procedures on their cost reimbursable contracts, because they have the attitude that recovery of the major portion of the additional cost is usually accomplished. One very easily incorporated discipline under cost reimbursable contracts is to establish and maintain fixed price contract program disciplines when the job comes into the facility. In fact, many companies don't consider the differences in their contracts; they establish and maintain the same control process regardless of the type of contract involved.

In earlier chapters, we discussed the need for cost estimating accuracy and realism. A company simply cannot properly implement a pricing strategy (whatever it is) unless the cost base is factual and accurately reflects the best efforts of the estimators. You cannot set the right selling price when the cost base is suspect.

At this point, we must address a common problem seen at many levels of marketing—including management. Marketers want to win the job they've chased for months and perhaps even years.

But, as is often the case, the cost estimates and initial pricing show you cannot be competitive, or would be at the high end of the competitive window. The first reaction on the part of the marketer is to cut the price. The point here is that this is putting the cart before the horse. It brings you right back to the target selection process and the preparations for a win through the implementation of your win strategy. Was the target properly selected? Were your R&D support projects properly organized? Did you study the customer funding trail? Did you study your competitors? Are they better prepared to win (on price)? These questions are not asked at the pricing review meeting on the day before your proposal is to be submitted. Planning is essential!

THE BUY-IN

The buy-in is probably best described as a tactical pricing procedure to position a company for an award where pricing (the lowest) has been determined to be the single remaining evaluation factor to be weighed by the procurement agency. A buy-in tactic for certain procurements may be considered essential by some organizations. They will set their inbound cost proposal level at the lower end (or below) of the competitive window for the procurement. It is an unreported cost sharing decision for fixed price contracts—a decision to supplement contract funds to cover total job costs. It is an unreported, but soon to be realized, cost sharing decision for cost reimbursable contracts; assuming, of course, that the company is successful in later convincing the customer to fund the major portion of the overrun.

In my view, the term *buy-in* should be removed from the marketer's vocabulary. Once that pricing methodology permeates an organization, there can be a serious breakdown of the procedure for job cost estimating and the setting of realistic and achievable job pricing levels for submittal to a customer. However, the practice is widespread and there are probably many substantial reasons for a company to pursue that course of action for certain procurements. In any event, the practice of adjusting pricing levels to win should be a function of the company management staff responsible for final pricing reviews; personnel responsible for preparing job cost estimates should continue to provide the best possible estimates. Otherwise, management has no solid base upon which to develop its final pricing decision.

How do you combat the buy-in tendencies of the competitor? All of the factors your company might consider if it is inclined to pursue this course of action will also be considered by your competitor. A good example of the buy-in tactic is for the sole purpose of *blocking* a competitor from the job to reduce his competitiveness (performance and price) on some future procurement. For example, a competitor may have studied his technology base—and compared it to yours—and has thus determined that he must block you from a specific procurement in order to improve his own technology base or to achieve a level of equality in the marketplace segments you both address.

A buy-in tactic may also be employed to better position a company within the marketplace segment it has selected. Often, the primary reason for the buy-in decision is to make it more difficult for the competition to participate—that is, to make it more difficult for the competition to achieve technical performance equality (or superiority) in a selected segment of the marketplace. A buy-in for the advanced development phase, for example, can better position a company for the follow-on engineering development and production phases.

This discussion is intended to demonstrate that a buy-in philosophy is not uncommon in the high technology business sector. The practice, however, is not condoned by many reputable companies, nor does it go by unnoticed by the government's procurement experts.

What one company may consider a buy-in by a competitor may simply be the achievement of a better pricing structure. The competitor may have found the formula for achieving the win price through a whole series of technology base and internal cost control improvements. You may never know for sure.

THE BEST AND FINAL OFFER

In general procurement activities, no discussion would be complete without mention of the infamous practice of subjecting the competitors to the best and final offer process—the *bafo*. This is a process often employed to get the best (usually the lowest) price for the products and services described in the procurement documents. This activity will *level* the competitors—bring them into the competitive window from the standpoint of performance–cost–schedule—and then subject the managements of the competitors to a game of *liar's poker*. The lowest offer may not be the best one for the procuring activity or the company involved. Objectivity is diminished. Realism is pushed to the back-burner. Realizing they are close to a win, company managements may arbitrarily reduce a price by 5% or 10% without fully considering the total impact.

This is why I have steadfastly pointed to the need for the best possible cost estimating. Company management cannot participate in a procurement where the bafo game is played, without first having accumulated the best possible cost estimates. The bafo response must be based on factual data. And, if your factual data will not allow the 5% or 10% reduction—then stand pat! Conversely, however, many companies involved in procurements where the bafo is to be played will provide for the price reduction by submitting higher initial pricing.

If companies could ever get their acts together, they could put a stop to this insidious practice by refusing to participate in procurements where the best and final offer mechanism is to be employed. Or, they could simply agree beforehand that the initially submitted price would not be further reduced. That's highly unlikely; action must begin with the procurement regulations discussed earlier.

The best and final offer strategy is a poor substitute for detailed negotiation. In fact, negotiation strategy for a negotiated procurement includes the elements of the best and final offer process. The company's negotiators simply stall as long as they can, hoping that they can retain as much margin as possible. These are the real gut issues in a cost proposal review process. They require the best possible cost definition package so that your management team can make the necessary decisions regarding the final price which will result in a fully acceptable contract.

You must guard against the tendency to forsake all that has

gone on before and agree to a 5% or 10% reduction based on the theory that you are so close to a win. Why jeopardize it now with a poor decision regarding a few percentage points right off the top. That's bad business practice.

All of these points must form part of the total negotiation strategy for the job. The plan must be carefully formulated at the time the initial pricing proposal is submitted. By doing otherwise, you are allowing the decision process to become a *hip shot* at best.

TERMS AND CONDITIONS

Terms and conditions—the boilerplate of the government contract—are primarily the selected contract clauses the procuring office wants levied on your company to control the conduct of the program. These clauses are usually required by either statute or regulation, and are taken from the Defense Acquisition Regulation.

In the purely commercial business world and in many industry-to-industry business relationships where the government is the eventual user of the procured product or commodity, a simple purchase order may be the only contractual instrument between buyer and seller. It is most likely that in the vast majority of intraindustry transactions the peculiarities of the DAR are seldom encountered. This does not imply that the DAR can be ignored by all except prime contractors—and perhaps key (first tier) subcontractors. Prime contractors and their major subcontractors impose certain DAR requirements on all of their suppliers whenever the end-item (the finished product) is scheduled to enter government inventory.

Having a good general understanding of the DAR can be extremely helpful to the marketer and marketing manager. While the marketing manager may also be held responsible for contract negotiation and management in some smaller organizations, most companies have a contract administration or management group responsible for that segment of the business. Marketing personnel then only need to be sufficiently familiar with the primary DAR content. In my view, such an understanding is extremely helpful when examining new business opportunities where the basic features of the customer's procurement plan (type of contract, performance, inspection, acceptance, and so on) are already beginning to form.

Often, the early formation of the planned procurement plan will also guide the marketing staff in selecting the best targets of opportunity. The fixed price R&D job is just one case in point where thorough examinations by marketing should be completed during marketplace and opportunity evaluations.

AUDIT AND NEGOTIATION

I can't cover every possible pricing–negotiation problem (or opportunity) you will encounter. Case histories could fill a library, if they were documented by contractor and customer representatives. However, strategies and their implementation during actual negotiations are usually proprietary to organizations and they are reluctant to discuss them openly. That's not surprising. It is the general intent to keep the process competitive. Once the books are opened, we would no longer have the essential competitive environment and the whole system would become unglued.

My purpose here is to provide simple logic to guide you and thus improve the decision process. Several examples will be given which should enable you to pick out essential and repetitive elements, which can then be applied to all but the most unusual conditions.

There are many negotiation strategies one can employ. Attitude management, prenegotiation cost audit management, and learning all you can about your adversary across the negotiating table all play a vital role. Negotiators are human (even though other unflattering descriptions have been given of them from time to time) and they, for the most part, simply wish to achieve the best contract terms for their organization. Negotiations are not much more than each side making concessions until the opposing positions are brought to the consensus line.

You cannot negotiate unless you understand the content of

the package you are negotiating. When the negotiator is misinformed or does not understand the content, the settlement may result in cost–schedule–performance problems once the work begins.

Negotiants must know precisely what is negotiable and what is not. Many times, it is not just a matter of negotiation over price but also over the elements of the job (specifications, schedule, and other factors). When properly negotiated, this will yield the contract pricing that the negotiants eventually agree is mutually satisfactory. Prior to negotiations you should identify the cost drivers in your proposal. The procuring activity may be willing to negotiate a relaxation of specifications, schedule, or other factor to pave the way for improved overall cost performance. But, caution should be exercised. In a competitive procurement, if the procuring activity is willing to relax a certain requirement for you, he will do the same thing for your competitor. You should not ask for relaxations that would make it easier for your competitor to achieve that requirement and thus improve his negotiating position. Often, terms and conditions will form a major part of a negotiation when certain requirements of such terms and conditions are balanced more favorably toward the buyer. You should probably never enter negotiations without first having thoroughly reviewed terms and conditions for content, understanding, and impact on the conduct of the job once the contract is awarded. However, one can expect that skilled contract negotiators will be prepared for this aspect of the negotiation process. They probably better understand this part of the procurement than they do the inner workings of the high technology product called for in the schedule.

In many negotiations, your customer's negotiator may have already reviewed the results of a broad range of prenegotiation audits conducted by procurement and audit specialists. A pricing audit is the most common. The purpose of the audit is generally seen in such terms as *reasonable, allocable and allowable*, and *questionable*. Within the government sector, one need only remember that the contracting officer (usually the negotiator) will have studied the audit reports. He will be prepared to challenge your position if the audit report clearly reveals a discrepant posture.

Prudent companies do not fear the audit; they provide the best possible trail for the auditor to follow. Thus, negotiations are streamlined and the negotiants can agree on contract terms more quickly and usually to their mutual advantage.

OTHER POINTS TO CONSIDER

One point that is often confusing for the new marketer, or the marketer examining the government marketplace for the first time, is the difference between the RFP (Request for Proposal) and the RFQ (Request for Quotation). Believe me, even the most experienced marketing manager will stumble when asked to explain the difference. The acronyms are used as if they are interchangeable; they are not if you want to be precise.

According to the DAR, when you submit a response to an RFP, the government procuring office may, under certain conditions, accept your offer by issuing a notice of award. The RFQ is an invitation to submit your proposal and enter into negotiations. The RFQ response does not constitute an offer which can be accepted by the procuring office without further negotiations.

The IFB (Invitation for Bids) is a procurement beginning with a formal advertisement. All bidders compete on the same basis against a detailed requirement. The submitted bids are received sealed, opened at the specified time, and the job is awarded to the responsive and responsible bidder who has submitted a bid most advantageous to the government.

Two- and four-step formally advertised procurements are also employed. You progress to a subsequent step by passing the evaluation criteria for the previous step or steps.

IMPROVING CUSTOMER–COMPANY RELATIONSHIPS

An excellent way to keep abreast of the trends in procurement practices and problem–solution studies and activities is to become an active member of the industry associations which—among their many other member services—serve to guide companies through the bureaucratic maze. In addition, these associations have access to government decision-makers and are thus able to lobby for or against certain procurement practices and procedures.

The adversary relationship between buyer and seller is not good for either the buyer's or the seller's organization. If our dwindling industrial base is to become healthy again, one essential factor is streamlining the government's procurement methodology—but while retaining the competitive nature of the marketplace. It is no easy task; there are no quick and simple answers. However, dedicated buyers and sellers can close the breech. They must learn to work together to achieve this objective.

Company participation in organizations such as the National Industrial Security Association, Electronic Industries Association, American Defense Preparedness Association, National Contract Management Association (and many others) is strongly recommended. These organizations are recognized throughout government and industry for the outstanding services they provide. It is a good idea for all high technology companies to consider participation through their marketing and contracts personnel.

CHAPTER FOURTEEN

MANAGEMENT OF MARKETING ASSETS

The marketing staff represents the major asset for the marketing manager. The plan for any given time period is only a plan until it turns into work. People perform the work. This chapter will discuss many of the tools (funding) provided to the marketing manager and his staff for the performance of their task.

Two major accounts are available to the government contractor: the Bid and Proposal account (B&P) the Independent Research and Development account (IR&D). Funds for these accounts are derived from the overhead structure, and are the additional approved funds received on government contracts to sustain the activities which are so critical to the marketing effort.

Additionally, marketing assets include the total departmental budget for all other activities, and company discretionary funds for additional research and development. While the marketing function itself expends only a small percentage of the funds through its own activity, the control of the expenditures by other company organizations against selected tasks becomes, in most companies, a marketing department responsibility.

In this chapter, we will describe the various accounts, their origin and purpose, and how simple controls can be established to insure that maximum benefit may be derived. In large organizations where millions of dollars are available for the development of business, budget (expenditure) control can become a complex task. Inadequate control can result in the misapplication of funds and the loss of a program (a win) critical to company profitability and growth.

THE R&D ACCOUNT

Earlier, we reviewed the research and development activities within a company and noted how vital these programs are to the overall business development effort. For many companies, research and development funds are not placed under the control of the business development group. Rather, they are generally considered to be part of the often discretionary funds available to technical staffs to pursue their selected laboratory programs.

There can be no separation of the technical organization and the business development group if the research and development activity is to have purpose and make its essential contribution to company growth. Management of the technology base is the basic consideration and should include strong marketing support, guidance, and control. It is only through this cooperation effort that the research and development activity will address technology base need while also contributing to the new business acquisition process.

Defense contractors have a significant advantage here in that substantial research and development funds become available through the very process of receiving government-funded contract work. True, the government requires that research and development activities be somewhat structured around its perception of its needs. Furthermore, annual reports are required to describe the results of the activities undertaken by the contractor. However, for all of the government's monitoring of this activity, the contractor has considerable flexibility in determining how the money will be allocated and how the various research and development tasks are structured.

It is therefore essential that both the technical and business development organizations participate in the planning for and the execution of the total program. Marketing managers who relinquish this role and look to the technical (engineering) organization to structure their own programs may find that they are unprepared to compete for a particular new business target. *Seed money must be carefully controlled.* The marketing manager and his staff play a vital role in its control in any successful organization.

Since most chief executive officers measure their companies on the basis of various ratios—all of which should be familiar to the marketing manager—it would be wise to also examine the R&D to sales ratio for your operation during the past few years. Surveys seem to confirm that a good growth trend for a company reflects a higher and sustained application of research and development funding. One can argue that increased R&D spending is simply the result of increased sales.

I am inclined to believe that a properly structured and sustained R&D program will contribute substantially to growth over the long term. Obviously, increased sales release increased R&D funds in most company accounting systems. A favorable attitude regarding the research and development program is not too difficult to obtain within your company when sales and net income are up and the growth line is positive. When these trends reverse for the short term (and even in the long term), that attitude will change rapidly. And, when research and development funding is reduced or even cancelled in the high technology organization, the impact can be severe.

Small high technology organizations which have yet to derive the benefit of IR&D through their government contracts, should still have a system and plan for allocating some of their own resources (as well as contract dollars) to the research and development task. Throughout my studies, I have found that among small high technology organizations, research and development plays the major role in the development (growth) of the organization and remains the most significant investment strategy.

THE B&P ACCOUNT

As for R&D funding, defense contractors accrue most of their bid and proposal (B&P) funds through the process of receiving government-funded contract work. Negotiated levels form a part of the approved overhead structure. Usually, the business development group in high technology organizations controls the allocation and rate of expenditure of these funds. The government expects you to manage these funds wisely, since they are intended to be used to cover the cost of responding to new business opportunities provided by the government.

The manner in which these funds are allocated to new business opportunities throughout a budget year should also be a cooperative effort. A marketing manager cannot become so protective of the account that proposals covering budgeted new business targets are inadequate and do not perform their function. Expenditure requirements must be developed by all company disciplines in cooperation with the marketing manager and his staff. Once established and then released, the expenditure of funds must be controlled. And control, as with all accounts, does not mean reading the monthly summations provided by the accounting function. Control means evaluation of the work effort at the point of expenditure.

Spend your B&P budget dollars *wisely*—but spend them! Have a spending plan for the entire fiscal year. Track expenditures against plan so that during the last quarter, you'll have sufficient funds available to perform the tasks you have scheduled for this period.

Managers will often defer the allocation of a portion of their funds. These are considered as management reserves for application to tasks which weren't clearly visualized at the time of the initial budgeting process. Many managers will use these funds as protection against budget overruns. They steadfastly refuse to make allocations out of these reserves. Often, such reserves remain untouched until late in the fiscal year; then they are expended rapidly against marginal business opportunities for the sole purpose of showing good budget performance.

This is a good example of nonsense management that pervades

many marketing organizations. Adjusting the B&P budget allocation on a quarterly basis (or more often if necessary) is an excellent control mechanism. The uncertainties of the high technology marketplace somewhat dictate how you should control this budget. However, if your target selection process is valid, you will have fewer of these uncertainties to consider during a budget period.

A CONTROL SYSTEM

Figure 14-1 shows one method you can employ to visualize how you allocate your funds from various accounts. By extending the allocation columns to embrace an evaluation period for prior years (three, five, or even more), you should begin to see expenditure practices that can be extremely helpful in establishing current-year budgets. Using actuals to serve as baseline data for a new budget process is a very common practice. However, this practice is not common in marketing groups for R&D, B&P, and other accounts established for new business acquisition. Figure 14-1 can be extended to include orders received or sales derived

BUSINESS OPPORTUNITES	PERCENTAGE ALLOCATION		
	IR&D	B&P	OTHER FUNDS
BUDGETED TARGETS (CURRENT YEAR AND SUBSEQUENT YEAR)			
BUDGET PROTECTION TARGETS (CURRENT YEAR AND SUBSEQUENT YEAR)			
OTHER TARGETS OF INTEREST (NEAR TERM)			
OTHER TARGETS OF INTEREST (LONG TERM)			

Figure 14-1. Asset allocation to business opportunities

from these orders in each of the new business target categories. Accumulation and study of your performance data can serve many useful purposes beyond the establishment of current-year budgets. Figure 14-1, when fully extended, may reveal an entirely new understanding of how well you are managing your two most vital marketing accounts—R&D and B&P.

THE BUDGET PROCESS

No discussion of marketing department assets would be complete without reviewing several of the disciplines associated with the departmental budget. Beyond general administrative and salary expenses, marketing managers face a broad array of other expense items to cover subaccounts such as travel, entertainment, acquisition of marketing data, consultant fees, and the marketing communications program. For many high technology marketing organizations, marketing communications by itself can exceed budget totals for all other subaccounts combined. Marketing communications may include advertising, trade show participations, product literature, product demonstrations, technical articles for trade journals, and papers for presentations to professional organizations.

Some managers still employ a zero-base budgeting process, where each item to be budgeted is initially viewed as having an equal draw on the available assets—the total dollar assets identified in the budget objectives. The technique generates a high volume of paperwork, but the quality of the budget doesn't appear to improve substantially. In this regard, it is interesting to note that even our federal government no longer employs this discipline. The Office of Management and Budget has rescinded Circular A-115, which required that the process be employed by all federal agencies.

Marketing department budgets should be established on the basis of the total company business strategy for the budget year and for the number of years contained in any business plan for the future. Marketing's contribution to successful company operations for the budget year is keyed to the capture of the budgeted new business opportunities. Additionally, a marketing budget for the current year must include those assets required to develop the long-term new business targets. To some extent, the budget can be divided into current-year activity and subsequent-year activity. As noted in earlier chapters, marketing activity for certain new business targets may embrace a three-year period, a five-year period, or even longer.

Your budgeting process may be over-simplified by such directives as "add 10% to last year's budget" or "reduce last year's budget by 10%." These are classic examples of management traps set by uninformed executives. Their directives are based solely on a sales projection and a cast-in-concrete net income line. Whatever departmental requirements may exist are adjusted as necessary to fit between these extremes. Companies developing their operating budgets on such flimsy factors are usually without a business plan or strategy for the current year or any other year.

Budget directions are more easily understood when given in the context of total business objectives, total business strategies, and various programs that have been developed for the implementation of these strategies. Then, when total sales are computed for the budget period (which includes all of the activities required to generate these sales), budgeting becomes a company-wide cooperative effort. The guidelines are realistic, and so are the budget projections that are formulated and submitted.

Many companies are utilizing a budgeting concept that allows for the cyclical, and perhaps unpredictable, nature of their business. It is not unusual for high technology businesses to utilize some form of flexible budgeting that provides for increases in certain operating budgets when demanded by a positive shift in the marketplace. Likewise, a negative shift in the marketplace can cause these same budgets to be reduced to gain further control over nonessential expenditures. Fixed budgets may prevent or discourage managers from addressing a new advantage in the marketplace.

The question for marketing managers is whether they will be tied to a fixed budget and miss opportunities or allow their budgets to flex as the marketplace demands. Control mechanisms are more severe for the flexible budget; otherwise, the undisciplined manager will use the flexibility to expand a market thrust toward poorly selected targets of opportunity.

Management of your marketing assets should not require a substantial amount of time and effort. By establishing a good budgeting discipline and a control system that will provide early and meaningful spending trends, you only need to monitor (assess) conditions on a periodic basis.

The major and vital task is the allocation process. Budgeted new business targets may not be converted to orders unless adequate resources are applied when they are required. This is the essential role of the marketing manager.

CHAPTER FIFTEEN

CONTROL SYSTEMS

To many managers, control of the marketing function equates to paper volume. However, the successful manager will tell you that paper does not control an operation; people control an operation. High technology marketing personnel are unanimous in their dislike of formal departmental and control systems. But, they readily accept the disciplines dictated by the customer and will devote a great deal of energy to fulfill presentation and proposal requirements.

In this chapter, our discussion focuses on the essential control mechanism for the marketing function, a basic format suggestion based on a broad survey of industry practice, and a description of how a simple yet proven system can provide control and total job (task) visibility to broaden communications between manager and subordinate—and throughout the total company.

A CASE FOR A CONTROL SYSTEM

Throughout my nearly three decades of involvement in the high technology defense business, I have encountered general dissatisfaction among marketing managers and their staffs about the never-ending requirements for paperwork. They believe, and rightfully so, that marketers should be out in the marketplace performing the basic tasks they were hired to perform. They also believe that any significant encroachment on their work and leisure schedules in order to deal with paperwork is counterproductive. Paperwork serves no useful purpose. Wrong!

Certain systems, procedures, and controls are essential. The problem lies not in the need, but rather in the development and implementation of inefficient and poorly structured and managed paperwork systems. Control is essential. What is also essential is a paperwork system that responds to need (both within the marketing or business development department and throughout the company) and requires only a minimum application of time and effort for its maintenance.

Immediately, we should discount the somewhat standard elements of the system that relate to travel and other expenses, performance appraisal, and certain other mandatory reporting and administrative requirements. Consequently, what remains are the essential and critical elements related to the acquisition of the budgeted new business targets.

Documentation requirements (paperwork) at this level should be structured around the total company business plan. (At this point, it is assumed that you have a business plan—a total company-wide plan describing your mission, business objectives, total strategies for the current year and the future, and specific programs for the implementation of these strategies.) Marketers who parade around in the marketplace performing tasks of their own choice do not contribute much to the new business acquisition process. They should be required to follow the script inherent in a business plan. And thus, as manager of the activity, you should implement control and reporting systems supportive of that plan—nothing more, nothing less.

A company business plan already reflects the conditions of the marketplace, identifies the targets of opportunity, defines the technology base, and presents a rather detailed assessment of current conditions and a projection for the company for the near and long term. At least, this is generally what a business plan is expected to provide, and those I have reviewed appear to have included these basic essentials. If your company follows such a procedure and considers its business plan (usually developed on an annual basis) as a *living* document, then you have your most critical control element firmly anchored. Further control mechanisms or systems now become greatly simplified.

THE BASIC CONTROLS

Figure 15-1 shows, in an over-simplified diagram, the basic control system for the total operation of the marketing (new business acquisition) process. This simple system provides solid guidance; virtually eliminates nonessential marketing activity; and provides the total company organization with guidance, planning information, program scheduling details, and accurate and timely status information. The diagram shows how a new business acquisition plan can serve as the central control system operating directly within the company-marketplace interface.

In reviewing many marketing functions and the control systems they employ, all seem to focus on some centralized mechanism for the new business acquisition process. Some are formal. Some are informal. Some function with great efficiency even when the system is not documented or even recognized as a system. However, whatever your motivation, I believe you can achieve a truly systematic and controlled mechanism for acquiring new business. You can better fulfill the requirements placed upon you and your marketing organization by the company business plan. You may even become more successful! (And, for the marketing manager, success is measured in terms of company growth and profitability, which are inseparable.)

| THE COMPANY BUSINESS PLAN | ↔ | THE NEW BUSINESS ACQUISITION PLAN | ↔ | THE MARKETPLACE |

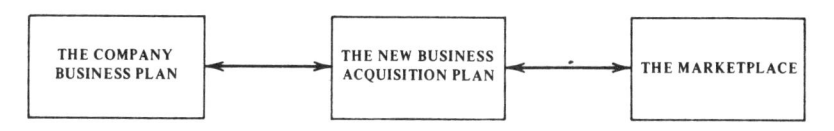

Figure 15-1. Interface for the new business acquisition plan

At this point, I want to give my personal observations regarding planning documents for your company and marketing operations. First, a company business plan—if it is to serve any useful purpose—must accurately reflect current conditions and quantify the direction of the business for the current year and the long term. As noted earlier, a company business plan must be a living document; it serves as the pivotal element in the total operation of the company's business. Planning documents prepared in response to a directive, then filed in the bottom drawer of some remote cabinet, serve no purpose. They waste your resources. If your business plan is regarded as a living document, then you should generate and maintain the interface document portrayed in Figure 15-1. The New Business Acquisition Plan and the company business plan work together as a control system.

Many organizations refer to this interface document as a marketing plan. This title is misleading. Marketing is a company-wide responsibility; business development is a company-wide responsibility. The problem lies in the title alone. If marketing is an identifiable organizational entity in your company, any plan generated by the group is seen as being for their exclusive use. This is not so. If properly structured, it is the implementing plan for the targets of opportunity depicted in the company business plan. It serves as the very heart and soul of the new business acquisition process. Thus, we can look to our marketing plan as the new business acquisition plan. And, we have now essentially described the only formal control system you may require.

While I have long been an advocate of the marketing plan discipline for the technical marketer—to exert a further level of control over his marketing activity—I believe this is no more than a subsystem for the acquisition plan. There is a major difference. An aquisition plan for a huge new target opportunity (a ship, an airplane,a tank, a missile system, a new command and control system) may include hundreds of elements. An acquisition plan for the widget we used as an example earlier may require only a few dozen pages. However, the basic content requirements and control aspects are the same.

To be effective, new business acquisition planning (and the

documentation which becomes a part of the plan) embraces every single discipline employed by a company to secure the contract award for the target being pursued. It is more than a how-to document. It is the core for every segment of the win strategy. It is obviously a highly proprietary document for the company, since it describes in detail how the company intends to capture new business, includes both customer and competitor assessment details, and describes all of the elements of the pre-proposal, proposal, and post-award program phases.

There are several reasons for not including an outline for the contents of a new business acquisition plan. That's the easiest part of the entire process. You, as the marketing manager, have already been assigned budgeted targets of opportunity. You probably have the basic elements of a win strategy. You have already determined most of the key steps needed to achieve the contract award milestone. Document them in the form of an acquisition plan. Distribute the plan to all key participants throughout the company. Now, believe it or not, you have the core for the only significant control system you need in your marketing department.

Following the release of the basic acquisition plan, all company effort related to the acquisition of that new business opportunity is keyed to plan requirements. All activity in the marketplace related to that target is conducted in accordance with the plan. Activity reports are prepared in the context of plan requirements. New and accurate information is worked into the plan as it becomes available. The marketing thrust is coordinated, purposeful, and reflects the key elements of the win strategy. You can control the movements of your marketers in the marketplace. Only one additional discipline is required.

Every plan should have *flexibility* to permit the insertion of new information. Periodically, you should review the acquisition plan to ensure that the latest information is included, and that the balance of the activity leading to a contract award milestone is again properly structured.

If you have ever had the feeling that company personnel are not all pulling in the same direction relative to a budgeted target opportunity, then I strongly suggest you adopt the acquisition plan concept. Tie their efforts to the plan and then watch for the surprising results. This is control. This is a control system that works. Discard all of the other control systems you may have implemented. They will probably no longer serve a useful purpose.

CHAPTER SIXTEEN

MARKETING DATA ACQUISITION AND CONTROL

The acquisition of marketing data is a basic requirement for the new business development organization. Once the data have been acquired, the responsibility for proper use in the decision process for new business acquisition rests primarily with the marketing manager. The basic elements are, of course, how much, best source, efficient study and use, and library control.

There may be hundreds of data sources available to the manager and his staff. Much of the data available is redundant. Marketing organizations all draw raw data from the same sources. The problem becomes one of acquiring what is required, in the format best suited to the company's need, for an expenditure in man-hours and material dollars consistent with that need and purpose.

Our discussion here relates to marketing data in the context of high technology and the government business sector. What are marketing data needs? What does the term *marketing data* mean to high technology marketers? Is the general market information available useful or does it require further study and analysis? Are general market data sources sufficiently reliable to support the company mission, objective, and business planning activity? Should the manager conduct his own market survey and analysis? Is the government marketplace sufficiently stable to serve as the basis for a long-term company business plan that is to be *cast in concrete?*

A GUIDE FOR DATA ACQUISITION

The decision process in the high technology business sector requires a solid data base. Acquisition, comprehension, and conclusions regarding applicability to a business plan serve as a baseline from which the marketing executive will generate recommendations for the direction of the company's future business. Unless certain disciplines are employed, the process will yield little useful information. The starting point is not just the acquisition of masses of data. The aquisition process should become highly selective—reflecting the need of the company's mission, objective, and strategic plan.

It has been my experience that high technology business environments yield much more data than can be properly assimilated by the marketing manager. The data flow within a company organization is often so rapid and complex by itself that little time remains to react to the meaningful and useful content of data flow from external sources. For many organizations, and for marketing managers, it has become necessary to segregate data acquisition and analysis into the three basic categories described throughout this text: company, customer, and competitor.

The previous suggestion that intramarketing reporting should be based on the outline format inherent in the new business acquisition plan, should also be considered for related reports generated within the total company organization. When reports are structured in this style, information exchange is more rapid and nonessential information has already been discarded by the writer of the report. It is very distracting for a busy manager when he is forced to wade through dozens of pages of reports to find useful bits of information.

Some day I would like to see a reporting system that requires information to be placed under one or the other of these two headings—*good stuff* and *garbage*. Obviously, this is an oversimplification of a continuing problem within most companies—large and small. Small companies can often reduce report requirements to simple monthly summary reports. Large companies—and especially those with widely scattered divisions and management organizations—have a much more stringent reporting discipline. It seems that administrative managers continually struggle with the problem of reducing their paper flow; some are more successful than others.

Many report requirements can be reduced to a standard yet simple format. It has been said that the best format ever devised is a blank piece of paper. Unfortunately, when a report writer uses that format, he usually gives the reader an overdose of garbage.

While internal reports are a significant part of data flow to the marketing manager, the most vital data flow is from external sources. Even for established and successful high technology organizations, data sources shift constantly. This is mostly a function of the marketing thrust—the segments of the marketplace currently being addressed by the company's new business acquisition process. Of course, there are many constant data sources utilized because of the company's sector. Most marketing managers have selected, and utilize, specific data sources because, over the long term, they have proven that they can provide essential information for the company's decision process.

Your primary data acquisition and control system should cover the government's acquisition process. There are hundreds of documents, directives, policies, procedures, circulars, guides, and planning details to describe how the marketplace functions. Documents of every conceivable description are published and available. They cover activities throughout the executive, legislative, and judicial branches of the government. How much of this mass of data you will want to obtain and include in your library is a matter only you can determine. Filling rows of file cabinets with copies of documents may convince you that your collection process is working. However, unless these data are properly organized to serve your essential business need, you may be better off by acquiring only what you need to support each individual new business target opportunity. Obviously, much of the government's documentation is general in nature and your library should contain these basic documents for whatever broad purpose they serve. It is essential though that you establish a discipline to protect the file against obsolescence. Changes in government policy and procedure occur on an almost daily basis.

THE DATA SERVICE ORGANIZATION

A question remains regarding the cost-effectiveness of establishing your own library and performing your own analyses. There are dozens of organizations whose sole business mission is the accumulation of a data base for all government procurement, from which you can select and order your specific and required data. These organizations provide an essential and continuing service for the small high technology business organization. For many, it is their only reliable data source. Large businesses augment their own data collection process with specific market studies and other reports.

In selecting service organizations to provide additional market data and reports, you should evaluate several of the leading companies. While most do a credible job, your specific needs must guide you to the best source.

CLASSIFIED DATA

Much of the essential data you will need may be controlled by government agencies because of security classifications. Acquisition then becomes a more disciplined process. The flow of classified information between government and industry is an essential part of the high technology business environment. (I will review some of the basic guidelines later.) Most high technology business organizations serving this market have already obtained facility clearances (for classified material storage and control) and clearances for their key personnel. For the company desiring to expand its business base, entry into the classified segments of the marketplace becomes an essential part of business strategy. The absence of facility and personnel security clearances restricts marketplace activity to a few very narrow segments. Expansion of the marketing activity, and thus the company business base, into classified areas requires a new and rigid data acquisition and control system. Growth for almost all high technology organizations is seen in this market segment.

BEGINNING A DATA SYSTEM

For the marketing manager who has yet to acquire his data base, there are some very basic initial acquisition procedures he should consider. Let's examine this start-up process.

I will assume at the outset that you already have an established business mission and product mix and, because of company strategic planning activity, have determined that expansion into the high technology government business environment is the next logical step. Whether you enter this marketplace as a prime contractor or at the subcontracting level, the basic needs of your library are essentially the same. You can begin by developing your comprehension of how the government is organized. The departments and agencies within the executive branch of the government will be your primary customers. Hundreds of documents are available from these departments and agencies that describe how they are organized, how they conduct their business, what they procure, and the procedures they follow. The legislative branch is of vital importance to high technology business because target programs pass through this segment of government for final approval and funding.

Where do you start? Obtain a copy of the *United States Government Manual*. Published annually by the Office of the Federal Register, General Services Administration, the manual is available to the general public through the U.S. Government Printing Office.

The importance of this manual is clearly indicated by the following extract from the foreword:

> As the official handbook of the Federal Government, the United States Government Manual provides comprehensive information on the agencies of the legislative, judicial, and executive branches. The Manual also includes information on quasi-official agencies, international organizations in which the United States participates, and boards, committees, and commissions.
>
> A typical agency description includes a list of principal officials, an organization chart, a summary statement on the agency's purpose and role in the Government, a brief history of the agency, including its legislative or executive authority, a description of its programs and activities, and a "Sources of Information" section. This last section is particularly helpful, providing the addresses and telephone numbers for obtaining detailed information on consumer activities, contracts and grants, employment, publications, and many other areas of citizen interest.

Experienced marketing managers have been exposed to and have probably used the benefits afforded by the Freedom of Information Act (FOIA) in acquiring essential information for their libraries and to conduct their business. However, misuse of the policies and procedures of the Act is becoming more and more a matter of concern at all levels of government and throughout the private sector. The obvious compromise of our national security aside for the moment, one can find many, many examples of businesses using the Act to their own advantage. While this is not the proper forum to discuss the merits of the Act or to provide details of abuse, one has only to study the record to quickly determine that additional controls are essential. As you seek to expand your data acquisition process, the Act can help by providing for a free flow of information. However, when FOIA provisions are employed to gain a competitive advantage, then the general rules governing good business ethics need to be reexamined by all of us.

You may have already reached the conclusion that information flow into, within, and out of the marketing function is both essential and comprehensive. This is true! The basic problem, however, is one of restricting the flow of nonessential information. Good business decisions are vital to your success; an uncluttered flow of information is essential to the decision process. Because high technology organizations are confronted with a constantly changing business environment, information flow must be accurate and timely. Consequently, information management becomes an additional discipline for the marketing manager.

CHAPTER SEVENTEEN

A MARKETING MANAGEMENT INFORMATION SYSTEM

A majority of the marketing executives surveyed complain about the *paper mill* they have either inherited or have allowed to develop within their department and between marketing and other departments. Paper flow is essential; the flow of the *right* paper at the *right* time appears to be the best definition of this requirement.

Solutions will begin to form when the executive can identify the information sets he requires to perform his assigned task. What are his information needs to support the decision process related to all key issues? Is format overriding? Should the marketers on his staff diminish their marketing efforts simply to satisfy a complex reporting system? Can he rely on filtered (summary) data? Can new formats be developed to consolidate (summarize) data related to all key issues?

I will now describe a concept for marketing information management. This concept addresses the problem of having the right information available on a timely basis to support the decision process.

A CONCEPT FOR INFORMATION MANAGEMENT

While it is a function of the size of your business, the command center approach to marketing management can be a useful tool for maintaining communications up, down, and across a company organization. I am not suggesting that you should go out and procure a computer-controlled large screen display system for your office, but I do feel that a certain level of command center operation may be of significant benefit in monitoring progress in a marketing department. Having all pertinent and useful data at your disposal—on a daily basis—can be a tremendous asset in reaching timely and appropriate decisions regarding the programs you are tracking.

It is a question of style. What do you need to efficiently perform your assigned task? Perhaps you can control adequately by having a few key pieces of paper in a three-ring notebook at the corner of your desk. Perhaps you need a number of multiple drawer file cabinets. But, whatever method you choose, the *organization of the material* into a useful format remains a key issue. The one basic problem you must alleviate is constant reformatting of the same old data.

Most management information systems survive because they are welded together by the mass of paper that flows between information-absorbing and information-producing nodes. But, paper flow creates inefficiency.

Your primary objective then is to develop an acceptable management information system which addresses your specific needs. Put nothing into the system that does not contribute to (is essential for) the decision process. Your objective is to get to the *gut* issues in near real time. If your business is so large and complex that it requires the processing speed of a computer to provide the required and essential data flow, then have one installed tomorrow.

Earlier, I described an operational concept wherein marketing reports are somewhat restricted to inputting the data base for a new business acquisition plan. Relevant information is provided; all other information is given a subordinate position in the attention it will receive from both the report writer and all the addressees. The reason should be obvious. Your primary function is to secure (book) the new business targets in your budget. The information you receive should direct your attention to the most important segment of your task. Your visibility and decision process should not be cluttered with information you cannot use effectively. Therefore, it would seem logical that if you could segregate information flow into priority categories, you would probably have a much better opportunity to make the right decisions at the right time. Remember that good business decisions are most often the result of having the right information at the right time. But, how is that accomplished?

One concept accepted by many is referred to as *strategic scanning*. This concept deals generally with the practice of deliberately reviewing most if not all of the information, in whatever form, as it crosses your desk. The theory is that you will derive a much broader understanding of the business environment, and thus your decision process can operate from a more substantial data base. If you ignore the unfriendliest of all factors, time, the concept has some merit for the high technology marketer and for other sectors of management.

Broadening the information base is beneficial. However, maintaining current plan focus is much more essential.

There are several methods you can employ for establishing your management information system. You can segregate information into the three major categories of company, customer, and competitor. You can steer all information flow into and out of the framework of the new business acquisition plan we discussed earlier. You can establish various key categories under such headings as research, advance planning, market data and analysis, near- and long-term business objectives, current and future new business opportunities, current programs, and budgets. You can establish rigid formats and reporting guidelines. You can place various reports into categories such as priority, essential, routine, and even general information. You can do all of these things, but if the reporters—your marketing staff and all of the internal and external sources feeding you information—do not wholly subscribe to your system, you will again be the victim of information glut. You might just as well let the system, whatever it is, function under its own inherent set of rules. The best you can hope for then is that you will be able to intercept all of the information essential to the proper operation of your marketing thrust and decision process. And, believe it or not, many successful managers operate under just such a system.

Let us continue by asking one simple question. What is your specific function as marketing manager of your high technology business organization? I assume you will answer by saying that your function is to properly analyze the company's technology base, assess new business opportunities, select the best targets of opportunity, and complete whatever work effort is essential to convert budgeted new business targets into booked orders. Fine! That's probably not a bad description for the high technology marketing manager. Consequently, if that is how you visualize your function, your management information system should be established to provide timely and essential information to enable you to properly perform that function. It really doesn't matter what type of system you install.

In the final analysis then, I will draw your attention to the new business acquisition plan concept discussed earlier. If you decide to employ that vehicle for control purposes, I see no reason why you should not also structure your information flow and its management around the concept. You will use a baseline management document as the hub of your system. Every bit of information then flows through the plan. What is relevant remains as part of the update. What is irrelevant flows through and becomes the routine (low priority) portion of the information system. In this manner, *you obtain an automatic segregation of information.*

Most marketing managers have expressed their concerns about the information glut—this massive flow of information about the business and marketplace they serve. One normal reaction is to add staff to accumulate, catalog, and file the information. Retrieval is another matter. Unless the information can be reconstructed for use when it is needed, massive data files provide no useful purpose.

Managers must identify and understand the information they handle. And, they have to describe, in very precise terms, the content of the information they need to receive and the format to be utilized. The objective is simple. In the fast-paced high technology business environment, managers can be overtaken by events from both within the company and external segments. One manager has described his information management concept in very simple terms—*give me relevant information flow on a timely basis.*

CHAPTER EIGHTEEN

SECURITY

The subject of security is a major one for any marketing manager associated with high technology, and particularly when the government is among his customers. In this context, security is the protection of both government classified and company proprietary information from unauthorized disclosure.

In dealing with government classified information (confidential, secret, top secret), the rules and regulations that must be followed are precise; the penalties for unauthorized disclosure are rightfully severe. Personal and company controls need to be addressed fully in order to provide the essential protection.

The control of company proprietary data is essential to protect the competitive position. Sensitive information relating to technology advances, new business strategy and planning, and cost data are examples of the company data base which must be protected.

This chapter will probably serve best as a reminder to the marketing manager that security is a major element of his daily routine, for both government classified information and for information related to company operations. And, the acquisition of a data base for his organization which would include the proprietary data of the competitor is intolerable, unless necessary exchange agreements have been executed.

PROPRIETARY DATA

The protection of company proprietary data has become a major concern for all business sectors over recent years. As a result, elaborate systems are being installed to protect files, correspondence, and computer storage. Elaborate systems for protection are required because elaborate methods are being employed to compromise the data held by many business organizations. We have all read the horror stories of intercepted data transmission, compromised files, and unauthorized entry into computer data storage systems.

Marketing managers and the business acquisition organization generate much of a company's proprietary information, or need to have access to all such data generated within a company. Therefore, the manager should establish whatever rules are necessary to comply with the protection requirements described by company procedures, or in the absence of such procedures, incorporate the essential levels of protection on his own.

Most progressive companies have a system for the protection of their data. However, regardless of the system they employ, scoundrels from within and without the organization will seek to acquire the information for their own purposes. To covet another company's proprietary information is one side of the coin; to obtain that information either overtly or covertly may be the most severe compromise of business ethics—notwithstanding the legal ramifications of such an action.

In our earlier discussion concerning the competitor and the establishment of your dossier on him, we reviewed certain practices for acquiring nonproprietary information. There is one more procedure you should incorporate. I have deliberately withheld this information until this point because here we are bordering on the acquisition of certain proprietary information.

It is not unusual for companies serving the high technology business sector to work together as team members for a major procurement. And, it is not unusual for these same companies to become competitors for the very next procurement. They team when their technology bases complement one another; they become competitors when their technology bases lead them to the same new business targets. When you do team (or negotiate some other close working relationship), you can learn how other high technology organizations function in the marketplace, how they relate to the customer community, and how they approach certain facets of their new business ventures. The process is simple and effective.

Study these companies carefully when you work with them in any associated capacity, such as subcontractor, supplier, team member, or partner in a joint venture. Often, companies working together to win a major procurement are stronger as a team than they are individually because they complement each other. One normal part of any such working relationship is the written agreement to mutually protect proprietary data. But, even within the constraints of these essential and binding agreements, you can easily determine many of the techniques they employ to reach certain objectives.

If you are to learn from the working arrangement, then you must pay very close attention to the dialogue of their participants and the methodology they employ to reach certain decision nodes. Usually, how they perform on the current job is a reflection of their normal practices and the direction of their companies. You can learn much by watching, asking questions, and carefully listening to the way in which they answer your questions.

Some will consider this a violation of the written agreement. I can assure you that whatever you are doing is also being carefully analyzed. Your competitor is probably as interested in learning from you as you are in learning from him.

Certain types of company and customer proprietary information are extremely sensitive, particularly during the closing phases of participation in a major new competitive procurement. It is during this period that the marketing manager and all other company personnel should employ certain additional security precautions.

You will note that I have identified proprietary information held by your company as well as your customer. The covert acquisition of competitor data held by your customer is intolerable if a truly competitive environment is to be maintained. In my judgment, almost all marketing managers and their personnel would much prefer to win their next award knowing that an honest evaluation and selection process had been followed. But, unfortunately, there may be a few bad apples in the procurement barrel. Your only recourse is to continue to follow the rules—a win under any other condition is, in my view, only a short-term gain at best.

Since the key evaluation factor in most procurements is selling price, many companies restrict their final pricing formula to the few select individuals within the company responsible for that decision. After all, your final submitted price is the bottom line for what may have been months and even years of careful management of the new business target acquisition process. And, the much-maligned government procurement process, during which the customer narrows the competitive field and then requests the best and final offer, may by itself tend to excite the bad apples toward a compromise of acceptable business ethics.

SAFEGUARDING GOVERNMENT CLASSIFIED INFORMATION

High technology business organizations doing business with the government—and especially with the Department of Defense—will usually require facility and personnel clearances if they are to become effective in the marketplace. Facility clearances are required primarily for the purpose of maintaining an acceptable security control system. This includes the receipt, storage, and control of government classified information. Additionally, facility clearances provide an initiating mechanism for obtaining clearances for personnel; and, on a need-to-know basis, acquiring government classified information related to the company's mission, business objective, and new business acquisition process.

There are perhaps hundreds of examples where information

classified by the government is probably no more classified than your daily newspaper. Government agencies often classify first and ask questions later. I happen to subscribe to the side of the issue that will impose restrictions wherever there is a possibility that an enemy would benefit through free access to the information. Obviously, there are some very valid arguments for a modified approach, since being overly restrictive may hamper innovative research and development activities throughout many high technology sectors. The point here is that much of the information essential to marketing success will be classified, and thus you should strive to obtain the proper facility and personnel clearances.

For those who may not have had the opportunity to examine the details of the government's industrial security system, I strongly suggest you obtain a copy of the *Industrial Security Manual for Safeguarding Classified Information* which is available through the Government Printing Office. The manual is issued under the directional authority of, and in accordance with, Department of Defense Directive 5220.22, Department of Defense Industrial Security Program. It establishes uniform security practices within industrial plants, educational institutions, and all organizations and facilities used by prime and subcontractors having classified information of the Department of Defense, certain other executive departments and agencies, or certain foreign governments.

In the high technology business sector, much of the information and many of the programs you will encounter will fall under one (or more) of the security classifications now in use. Security is normally (but not always) linked to end-item use, performance levels, operational capability, and deployment methodology. As one progresses up an assembly tree from the component part to the total system, the security levels become more and more stringent. The reason is obvious; system performance cannot always be determined by examining only the low level component parts *prior* to their end-item assemblage. The tendency is, of course, to relax security provisions for as many of the component parts as possible.

Marketing managers examining the marketplace for their targets of opportunity should be mindful of security provisions. It is not unusual to be attracted to some segment of the marketplace—because of the push of your technology base—only to discover that either your facility or you and your personnel do not have adequate clearances for access to the program information.

The manual described earlier also contains guidelines for securing proper clearances. It is important to note that facility and personnel clearances are not granted just on the basis of a simple request. The periods of time required for investigative agencies to perform their tasks are lengthy, particularly for secret and top secret clearances and for access to compartmented information. Having the proper clearances, however, certainly paves the way to a broader comprehension of the high technology marketplace, and provides for improved selection of targets of opportunity.

CHAPTER NINETEEN

PERSONNEL SELECTION, MANAGEMENT, AND COMMUNICATIONS

Every marketing manager is aware of the need for acquiring the best possible staff to conduct the activities of his department. Acquisition of the right person for the job is only the beginning. How he manages his personnel will often mean the difference between success and failure for segments of his new business acquisition process. Usually, the personnel in high technology marketing functions are self-motivated, highly trained specialists who may regard the strong arm of management as a threat; and, in some instances, inconsequential to performance within their assigned areas of responsibility.

Significantly, a majority of high technology business development managers do not have the patience or are not given the time or resources to develop their personnel. They are much more inclined to bring skilled marketing professionals into their group because of pressures to achieve whatever new business acquisition objectives have been budgeted. Unfortunately, managers do not usually acquire the perfect staff, and thus additional personnel development is needed.

This chapter will examine several key managerial techniques and marketing group interactions, and activities that appear to be generally practiced by successful new business development managers.

A marketing manager said: "I hire skilled people, pay them a good salary, tell them what I expect, and then turn them loose in the marketplace. They either perform or I look around for replacements." The consensus strongly suggests that this is a manager headed for trouble. Fortunately, there aren't too many of these rascals in vital management positions.

The manager of the skilled high technology marketer has many advantages over other department managers in a company hierarchy. Often, he has only to perform certain critical personnel maintenance tasks. He is not burdened with baseline personnel development tasks. Nevertheless, there are several levels of personnel development that should not be overlooked.

PERSONNEL SELECTION

My survey of marketing executives in the high technology government marketing sector shows some interesting results regarding the levels of expertise they expect to see in candidates for positions in their departments. They ranked candidates as follows:

1. those with experience in the government marketplace
2. those with experience in program management or high-level engineering (design and development) assignments within the company's technology area
3. former government service employees (military or civilian) with experience in procurement, program and systems management, or service organizations
4. all others

It is interesting that these executives look for personnel who have already been exposed to the government marketplace and have learned, *through experience*, how to address some of the unique challenges of that market. It is obvious that most marketing executives need the immediate productivity of the experienced marketer; there are no programs to quickly train other candidates. And, it is also interesting to note that these marketing executives look for personnel who have training and experience within the technology sector addressed by the company's products.

It is simply a question of the ability to represent the technology properly—to work productively in the specific market segment selected by the company. But, above all, I believe it is a reflection of the levels of technology and marketplace experience held by managers of the government marketing function at most high technology industries. They reached their managerial positions through the process of promotion from within their organizations (from previous low level marketing positions or from the program management or engineering disciplines) or were acquired from a competitor where that same experience was gained. Then, thrust into a management position, they obviously look for personnel who can function in the marketplace without long periods of indoctrination and training, and with a minimum of supervision. These managers are predominantly technology-trained and marketplace-oriented, and may very well be at or near the bottom of the scale with respect to administrative and personnel management skills.

Marketing team members may not be full-time employees. In fact, many high technology organizations employ one or more part-time consultants to cover specific marketplace sectors or specific programs within a sector where special levels of expertise are required.

In recent years, the high technology government marketplace has become inundated with former government service employees—both civilian and military—serving as special consultants to industry. Many of them have outstanding credentials and broad contacts within their prior government organizations, which are very useful as they pursue new business acquisition interests for their new employers.

Most of these consultants are not high technology marketers in the truest sense of the term. They are more likely to be regarded only as *social* marketers. They open doors, complete essential introductions, and often notice the subtle changes occuring in the marketplace which would be overlooked by a marketer having no prior government connections. And, as is often the case in dealing with high technology procurements in the government marketplace, it is the unrecognized subtle change that may have the greatest impact on a company's business acquisition process.

Consultants, therefore, can and do play an important role for the marketing manager. They can make significant contributions to the acquistion of new business *if* they are carefully selected and utilized to the fullest advantage.

The primary objective is to assemble a marketing team whose collective energies will be directed toward the fulfillment of the company business plan. Innovation and creativity are expected and desirable, but only to the extent that approved business objectives are being addressed. Once the primary new business targets have been selected and become part of the total company commitment, the marketing team then has a clear-cut mission—capture the orders!

MANAGEMENT AND LEADERSHIP

For years, I have struggled to come up with a descriptive term that would be applicable for nearly every marketing manager confronted with the duality of purpose within a marketing group. I like *mission focus*. Once the needs of the company's business plan are fully understood, the mission of the marketing group becomes clear. Then, it becomes the task of the manager to focus group attention on that mission. In my view, it is the fine-tuning process to maintain focus that separates the successful marketing manager from all others. This process is essential for the high technology government marketplace.

I know of no other market segment which is as volatile. I know of no other market segment requiring so much carefully considered fine-tuning of the business acquisition process. To use the terms *always* and *never* to describe the high technology marketplace is to deprive oneself of the flexibility to adjust—to fine-tune—almost on a daily basis. This is the function of the marketing manager, and it may very well be his most important function.

Perhaps this is nothing more than a cursory examination of management by objectives. The objectives are clear-cut. They are spelled out in the approved company business plan and further analyzed and expanded by the new business acquisition plans discussed earlier. But, as Drucker has stated, management by objectives requires that such a program must be purposefully organized and be made the living law of the entire management group. Not too many high technology marketing organizations operate under such a disciplined structure. Thus, it is usually the successful marketing manager who, by his own initiative, will pull his marketers toward the new business acquisition objectives and then fine-tune their marketing activities, when necessary, to the degree required.

It appears to be universally accepted that a manager is expected to set objectives, organize work effort, motivate and communicate, measure the performance of the group and the individuals within the group, and develop himself and his people. What this listing does not explain is that unless the group plays a role in the setting of objectives, achievement will be much more difficult.

COMMUNICATIONS

How should the marketing manager manage? There are hundreds of experts (and as many more who strive for that title) who write and lecture on this subject. Most of us have worked under managers who we thought were good, bad, or in between. In the context of this book, it is my feeling that the most effective manager is the one who fully understands the technology and marketplace addressed by the company, who will require the marketing group to become an integral part of the total company business planning (and target selection) process, and who will then maintain an *open communications link* as the work moves forward toward the achievement of mutually established objectives.

Much of the success that people achieve comes from attitude. As a manager, you will make or break your marketing group by your own attitude. If you give them a sense of participation in the planning and implementation of decisions, you enrich their jobs and give them a feeling of involvement. This will then be reflected in the quality of their work, enthusiasm, dedication, and, more importantly, results. Communication plays the leading role in imparting this sense of direction or participation.

Managers must first know what it is they want to achieve so they can relay specific and continuing guidance to their marketers. They cannot allow them to parade around in the marketplace, each doing what they believe is best for the company. A coordinated effort is essential. And, you can best coordinate by communicating—by articulating the game plan in its clearest terms. When the marketing force participates in the formulation of the game plan, communicating the sustaining instructions becomes an easier task.

When you stop to consider that on an annual basis the average employee in your marketing group devotes only about 20–25% of his available time to the company, you may suddenly realize that he isn't a machine to be turned on Monday morning and then turned off on Friday afternoon. He may very well consider his job and the tasks you assign to him far less important than his family or other outside interests. Your sphere of influence, based on time alone, is reduced even more if you fail to communicate effectively during the short periods when he is available to you. In the marketing profession, where personnel are travelling so much of the time, your paths may cross only briefly. Thus, effective communication becomes an even more critical requirement.

If this is beginning to sound more and more like a lecture on participative management, then perhaps that is the real message I wish to impart. So much is now being written on the subjects of productivity, quality circles, and participative management theory and practice. But, in my view, it's nothing new. The objectives are quite simple. When workers understand company objectives—having been given the opportunity to share in their development—they more willingly and eagerly participate in the work effort required to meet these objectives. And, this concept applies totally to the company's marketing group. The marketing manager who applies these principles to the operation of his group is well on his way to meeting whatever objectives are summarized in a company business plan.

High technology government marketing may be a different game for some. Many marketing managers won't dirty their hands with the organizational and control aspects of their job. They would rather be the *wheeler-dealer* who makes things happen because of who they know or who they can influence. And, while the government marketplace is besieged on a daily basis by thousands of high technology marketers, there are specific disciplines that cannot be overlooked if a company wants to earn the best reputation or build upon (or even restructure) one which has already been established.

CHAPTER TWENTY

MARKETING COMMUNICATIONS

Over half of the high technology marketing managers surveyed could not verbalize their marketing communications concept for either the new business acquisition process or their company. It is my impression that very few organizations maintain a totally integrated marketing communications program; instead, the elements of an ill-conceived program are scattered throughout the company. There would be no easy way to measure effectiveness.

Marketing communications embraces a wide range of disciplines, including advertising and public relations, product literature, trade shows, proposals, special reports and other communications, and every other contact with the marketplace, regardless of method or timing. We'll explore this subject from the viewpoint of the high technology government marketing manager. We'll determine how he should evaluate his current practices and develop an overall plan which can become an integral part of company short- and long-term strategy.

Marketing communications costs are often viewed as a suspect departmental budget line item, and thus subject to reduction and even elimination during a business downturn. This and other inappropriate practices need to be reviewed in order to show how an effective program can be developed and scaled to the size of the business, the growth plan, and the available assets.

THE PERCEPTION

Every bit of information flowing from your company into the marketplace must be considered under the umbrella of marketing communications. This embraces the activities of your CEO, as well as the technician working on a product application problem with an isolated user agency—and everything in between. Every single company employee who has any contact whatsoever with the marketplace is contributing to the perception the customer and marketplace have of your company.

Consequently, it makes very little sense then to allow this perception to develop or *not* develop on a random basis. It is necessary to establish a marketing communications program to create and sustain a desirable perception.

If you presently do not have a good understanding of the perception of your company in the marketplace (in this context, marketplace is the total environment outside the company), then I suggest you initiate a very simple but effective program to learn what it is. During the course of your employee contacts in the marketplace, have them ask one simple question —*What is your perception of my company?*

For purposes of obtaining a good cross-section of response, you should divide the marketplace into categories. For example, your marketers can obtain input from their customers (both new and established). Then, the personnel recruiting manager or director for your company can obtain input from new recruits and potential employees during the interview process. Ask the question of any competitors with whom you may have contact. Ask your advertising and public relations manager (or agency personnel) to provide input. Ask company employees. Ask the man on the street. Ask the business editor of your local newspaper.

The data base accumulated through such an exercise can be enlightening as well as frightening. You'll soon learn whether your marketing communications program is effective, partially effective, or ineffective in every sense.

I happen to believe that this is a good first step to take if you are in the process of establishing a marketing communications program for the first time, or even if you just want to determine the effectiveness of the one you already have in place. While such an activity may not be too dependable (that is a function of the people you have contacted), it will nonetheless give you a good indication of how well your current program functions and the motivation to make some essential improvements.

The purpose of your marketing communications program is simply to mold the perception the public has of your company. Every marketing manager can tell his own stories about certain competitors—the perception he has of the competitor which in turn is usually based on the perception the marketplace has of him. Bad news travels fast. And, bad or uncomplimentary publicity is remembered longer than the good publicity a company can generate. The following are heard often: "good designs, but poor delivery"; "unreliable products"; "excellent technical organization, but high priced"; "outstanding proposals, but poor performance after contract award." I could add dozens more, but the point is simple. If the perception of your company in the marketplace is not the perception that is essential to your new business acquisition process, then change it!

THE ELEMENTS OF MARKETING COMMUNICATIONS

At this point, it is important that I specifically identify the elements of marketing communications. Some of them may not be part of your program. However, keep in mind that every communication emanating from your company (and likely to be heard, read, or interpreted in the marketplace) should be considered part of the program. Obviously, many of the elements listed play only a minor role; but they are important. The problem, of course, is that it is difficult, if not impossible, to measure the individual contribution of each one.

- advertising
- product and other literature
- technical and other professional papers
- trade shows and exhibits
- trade association participation
- no-cost directory listings
- paid directory listings
- public relations
- public affairs and employee participation
- company philanthropy
- internal publications
- press releases (any facet of the business)
- employee communications programs
- presentations (formal and informal)
- logos and trademarks

As you can see, I have included the widest possible range of internal and external programs under the banner of marketing communications. This is intentional. I believe that any form of communication from any segment of the company—if it will eventually reach the marketplace—must be included. The reason is simple; you are shaping the perception people have of your company. This is the real purpose of a marketing communications program.

TECHNICAL PAPERS AND PRESENTATIONS

The preparation of technical papers for publication and presentation should be a key element of your marketing communications program. Almost all progressive high technology organizations use this vehicle for the purpose of disseminating information. It is an inexpensive method of reaching selected audiences for the purpose of building technical reputation and guiding the perception they have of your company.

The editors of hundreds of trade publications are constantly on the alert for good technical articles. There is probably no better way to promote the company technology base and new accomplishments. But, the marketing manager will often encounter a problem in the development of papers and presentations to complement the current marketing thrust.

Engineers and technicians have a tendency to write about what they know best. Consequently, the technical article or presentation will not always serve the current business development or new business acquisition thrust. Instead, many articles will only minimally contribute to the building of the desired perception in the marketplace. However, they are essential and should not be discouraged.

Many companies award honorariums for papers which are published or presented. Often, the opportunity to earn additional income (along with the prestige) is a sufficient incentive for the busy engineer.

Marketing managers can play an important role in the professional paper program. By helping to select the subjects to be covered, the manager can supplement his direct marketing program and other marketing communications activities based on his perceived need in the marketplace. Well-written and informative articles and presentations can supplement technical and management proposals, serve as introductory material for a new customer, and act as a reminder for a source selection board that your company is the best among all the competitors.

MARKETING COMMUNICATIONS AND BUSINESS PLANNING

Marketing communications programs are usually not given much space in the company business planning process or the business plan itself. In my opinion, this is a very serious mistake. A company business plan is usually regarded as the plan for the future of the company, and includes the objectives and strategies deemed essential to achieve the short- and long-term business mission. And, since the company CEO is the primary architect for the business plan function and the plan itself, whatever one reads in the business plan is a reflection of the CEO. Thus, if marketing communications are not included in the plan, then the CEO doesn't consider the program essential. What other conclusion would you choose to draw?

In his excellent series published in *Aviation Week & Space Technology*, James Pierce, the publisher, ends each article with the reminder that "productive advertising starts at the top." What Pierce is saying for advertising applies to all marketing communications disciplines. If your CEO does not subscribe to an integrated program—let alone the advertising segment—it becomes an element of secondary importance throughout the company.

If you believe in a good marketing communications program—that is, if you believe such a program will contribute to the new business acquisition process—then I strongly suggest you get your CEO in the boat with you. His support is essential; otherwise, your program will surely fail.

The concept of including marketing communications in a company business plan is not new, nor is it cumbersome. Company business planning and the business plan itself will usually detail programs for the implementation of strategies. I can think of no better place to describe the role of marketing communications and the plans and procedures for augmenting the other programs. An advertising plan can be included; it can be tied directly to the new business targets described in the business plan. The plan for technical articles and presentations can be keyed to important milestones for each and every program. Trade shows and exhibits can be tied to these same milestones. Each of the 15 elements of a marketing communications program I have already listed can be treated in a similar manner. The objective is to show interrelationships, and how marketing communications can also play a vital role. Consequently, like all of the other elements of a company business plan, marketing communications will be given a share of the attention. This is progress!

ADVERTISING AND THE HIGH TECHNOLOGY GOVERNMENT MARKETPLACE

"To advertise is to get them to buy." With this bit of free advice, I could leave the subject in a precarious position and let you decide what it means and how you should proceed with implementation. But, I won't.

In high technology business, you can find many different concepts regarding advertising. One concept embraced by many leading public relations and advertising experts is that of *image* advertising. The intent is to continuously tell the buying sectors (and others as well) about technological accomplishment, philanthropy, concern for people, public issues, and other image-building themes. These are not scatter shots at the general public or general marketplace. They are carefully planned and developed programs to implant specific messages at specific points in time.

Marketing managers and others within the company responsible for advertising should study the print media carefully. Outstanding examples of image advertising are presented by Hughes, United Technologies, Martin Marietta, Sanders, and others. The advertising is not truly product-oriented. However, these companies will, in time, *hook* a reader so that when their advertising appears later to promote a specific product or a new accomplishment, he will obediently stop, look, and read. Thus, the primary objective was accomplished. And, this is also an excellent example of *continuity* in advertising.

This is not continuity in the sense that the same old message is forced on the reader at every opportunity. Rather, it is continuity in the sense that identification of the *company* is sustained; and in the sense that it reflects a plan for advertising—a plan for marketing communications.

In this book, I want to formulate an advertising plan for the general supplier of high technology products for the government. Advertising in this category is designed to supplement all other facets of the new business acquisition process—from the direct marketer contact to the submittal of the final cost–technical–management proposal. In competitive government procurements, there are several distinct milestones in almost all programs which lend themselves to the further support advertising can provide.

Advertising plans for the government contractor need to be examined carefully, and with a full understanding of each of the new business targets to be addressed by the advertising program. There is no general rule to follow; instead, each program has its own variables that need to be considered separately. These variables cover differences in technology, program schedule, decision-level personnel in the customer area, user personnel, specific program issues, and the posture of major competitors. The specific message you want to implant in the minds of the customer's key personnel and the method you employ to carry that message are vital elements of the ad campaign.

A review of Figure 20-1 should clarify why effective advertising campaigns for the government contractor are difficult to organize and implement. Each of the segments shown require individual tailoring since program elements and message direction and content can vary widely. To be fully effective, you should initiate your advertising thrust (begin to tell your side of the story) soon after you have completed your target selection process. For those key targets earmarked for exploitation through advertising, you can begin with general ad content to acquaint (remind) key customer personnel that you intend to participate in the procurement and should be considered a viable competitor. Obviously, you don't identify the program or strive to answer questions which they (the customer) have yet to formulate. The thrust is simply to bring your capability, credentials, and technical expertise into the picture.

As you follow the line of progression for advertising, as shown in Figure 20-1, you will note the program segments which should be addressed. The message of your advertising is a function of the program segment, as in our example. Advertising is considered the heavy artillery in the arsenal for marketing communications. However, to be effective, the advertising program must be carefully developed to carry the right message to the right customer personnel on a timely basis.

One other aspect of advertising is vital—you must develop a listing of the key trade publications you will utilize for print

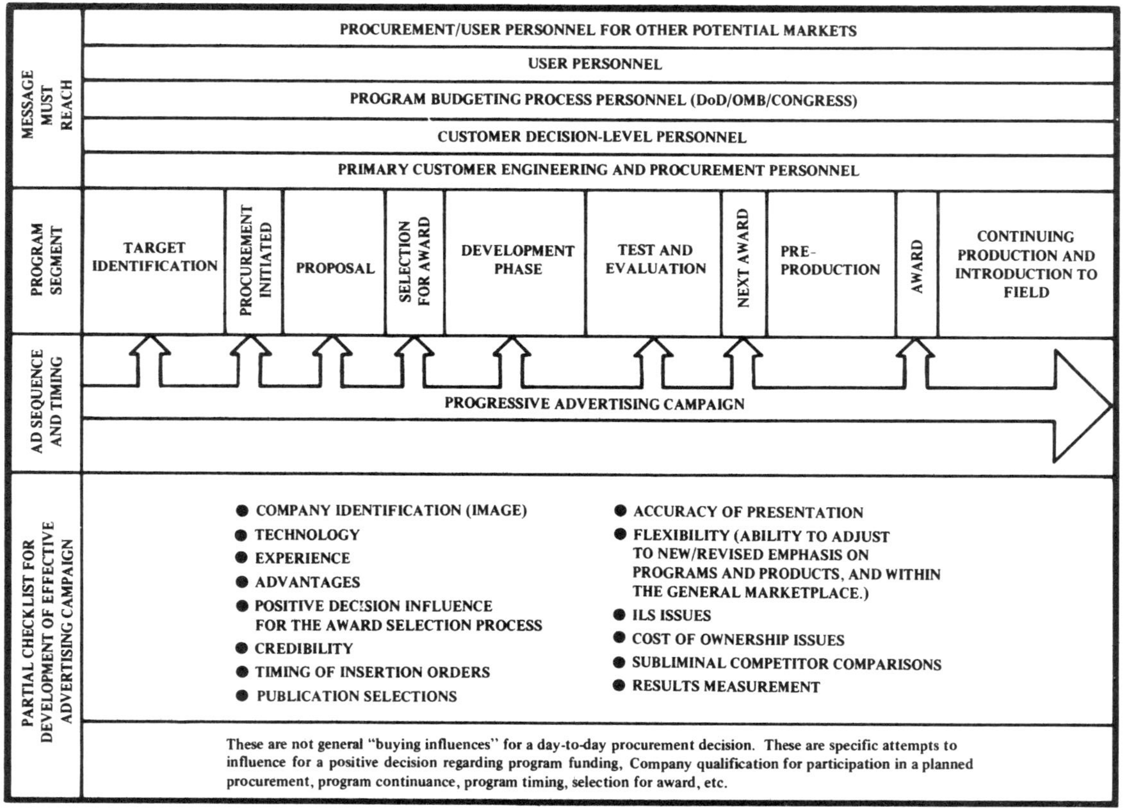

Figure 20–1. Progressive advertising campaign layout

advertising. In Figure 20-1, I have identified the key personnel who need to receive your advertising message. How do you reach them? The simple answer is that your advertising should appear in the publications they read. But, what do they read? Again, it is simple to find out. With an integrated marketing communications program locked to a new business acquisition plan, your marketers will pay closer attention to reader preferences. It is part of the marketing intelligence-gathering process. When your marketers are making personal calls on key customer personnel, they should note which publications are in their offices. If your customer contact uses a library service within his organization, then visit this facility to learn which publications are received and distributed. The process is simple and, in my view, absolutely essential.

Figure 20-2 is included to show one technique for organizing the listing of publications you have selected for ad placements. There are many variations of this format; you can develop your own to suit your particular needs. The objective is to identify the publications you will use out of the dozens and perhaps even hundreds you can identify. The final selection process is usually a function of budget limitations and *getting the most bang for the buck.*

THE BUDGET FOR MARKETING COMMUNICATIONS

The zero-based budgeting process can be applied here perhaps more advantageously than for any other sector of the business. The objective is not to compute some acceptable percentage of total sales or gross income and then look for elements of a communications program to fit the available dollars. Instead, you should examine the *need* first. What do you want to accomplish with your marketing communications program? What programs, products, and capabilities do you need to promote to aid in the acquisition of the new business targets in your budget? What are your key competitors doing?

For the high technology company whose primary customers are the Department of Defense and other government agencies,

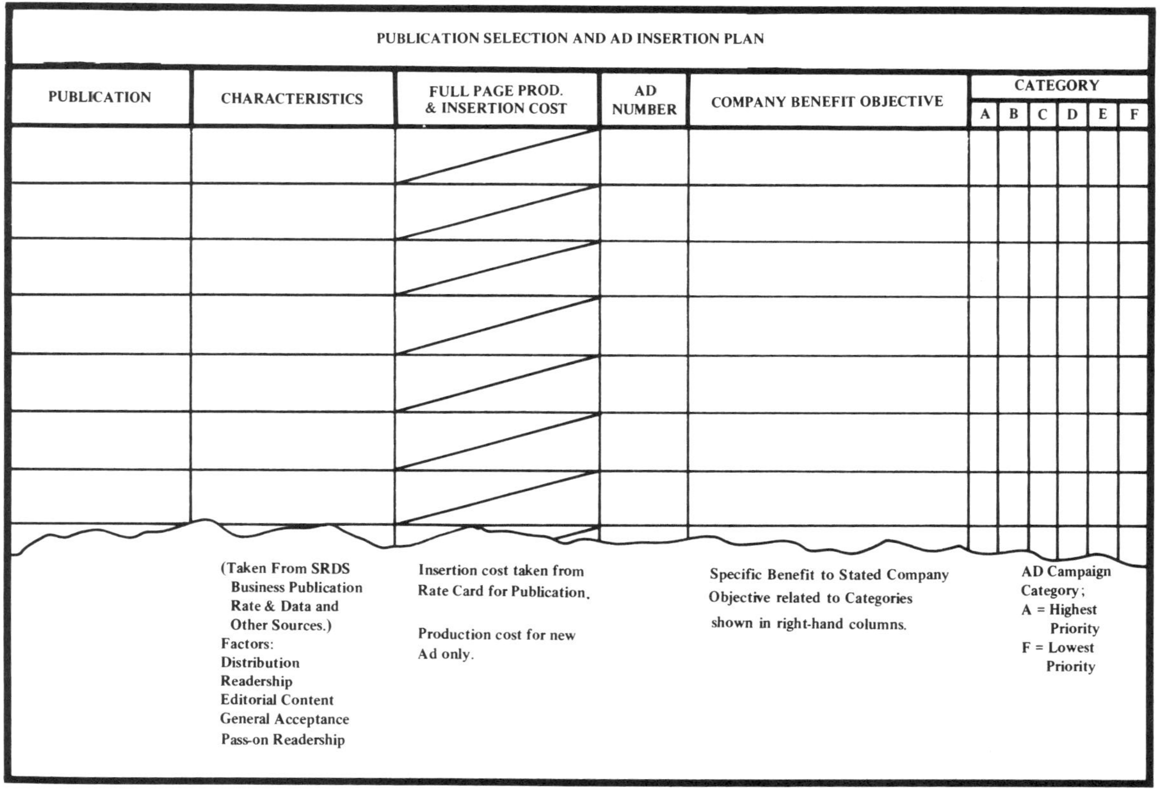

Figure 20–2. Publication selection and ad insertion plan

the objective is not so much to sell the existing product as it is to further influence decision-level personnel. This includes obtaining favorable consideration for the programs of interest to the company and for the company itself in a competitive procurement; countering the competitor's message, whatever it is; and promoting the company's capability, experience, and technology base.

Timing is a vital consideration. You must place your message before decision-level personnel while they are in the process of making decisions regarding your company and its capabilities. Advertising agency researchers will tell you that you should maintain an *effective presence* in the media—that is, repeating your message in selected trade publications as often as necessary in order to establish a continuity of information dissemination. This is why I have included the information used in Figure 20–1; satisfying an effective presence criterion must be weighed against key program milestones or events. Timing, in my view, is much more important than simply satisfying the presence criterion. If your placements are timed properly, you will automatically satisfy the requirement for presence.

This is also the case with all other elements of your marketing communications program. Trade show participation, release of technical papers, development and release of product literature, and all of the other elements of the marketing communications program should be timed to support the marketing (selling) thrust.

The budgeting process is improved by laying out your projected program across a 12-month period, for example, and then inserting the time and material dollars required to achieve the various elements of the program. You'll find that your budget requirements will usually be higher, and you may be required to trim various plans to fit some predetermined maximum allocations. However, I can assure you that if you believe in your program and can present a definitive plan and show how it will aid in the new business acquisition process, your requests for funds are much more likely to be approved. You should argue based on *need*—not on some industry norms for a percentage of sales or income.

THE COMMON DENOMINATOR

The guidelines provided for advertising and technical papers and presentations apply, either wholly or in part, to all other facets of your marketing communications program.

The listing provided earlier in this chapter may or may not include all of the elements of your program. But, whatever the elements in the listing for your operation are, there must be a coordinated activity to tie them to the purpose of your program and the objectives you have outlined.

The elements in your listing should be mutually supportive. Each one should carry your intended message or messages in its own style and format. But, when viewed as a whole, the intended theme or themes, and the direction in which they are pointed, should be clearly visible. Your total audience is complex. If your message is to be clearly understood in every sector of the marketplace, you must tailor it to fit each one specifically.

The common denominator for your marketing communications program is the purpose and objective statements which should appear in your company's business plan. From that foundation, each of the program elements can be addressed in a manner and style to fit the various marketplace segments. *Managed communications*—a strong weapon when loaded and aimed properly!

CHAPTER TWENTY-ONE

THE CHANGING MARKETPLACE

Throughout Chapter 10 ("Forecasting"), the general theme is learning how to use specific and available indicators to improve the new business forecasting process. Since we cannot predict the future, our next best bet is to speculate. Armed with the best evaluation of currently available indicators and trends, we can attempt to set a course for company, departmental, and personal objectives.

What should marketing managers strive to accomplish in the years ahead? Will the burst of newer technologies and the complexity of the marketplace change the marketing management role? Will the marketers in a new business development organization need to become more technology-oriented, or will the requirements of the business dictate greater emphasis on finance, operations, personnel and interrelationships, and planning and organization? Will the government continue to play a major role in the advancement of new technology or will industry, alone, assume the leadership role? These are very difficult questions to answer; and I suspect that simple and direct answers are not available. Each marketing manager must examine his own operation, his style of management, and his company's mission.

Advertising agency personnel continually talk and write about positioning your company—that is, what you want the readers of your ads to remember about you. But, positioning can have a much broader meaning. Positioning your company for its future in the high technology business sector is *your* primary job assignment. Company business plans, your marketing plans, and your new business acquisition plans should all reflect the approved objectives which have *positioned* the company to address the needs of the marketplace.

If your business is primarily aimed at the defense market, are you positioning your business around the perception of need as seen by your customer, or the need as you perceive it? Companies seldom dictate need, but they may assist in its formulation. Unless your company is in a position to evaluate threat (in terms of our national security), then you probably respond to need based on the customer's perception of it. As the threat changes, so does the need. Thus, flexibility is essential and your plan for the future should allow for change when it is necessary.

The very basic meaning of the term *positioning* leads me to perhaps the single most important management task you face as you look to the future. Throughout this text, I have stressed technology base management. It remains, in my view, the single most important requirement for the high technology business organization. Without a strong and adequate technology base for whatever mission and marketplace sector you address, you will fail. You will fail to increase market share in an expanding market and in achieving company growth and profitability objectives.

Marketing managers are quick to point out that they provide adequate funding for technology base maintenance and expansion throught the IR&D program and investment of company funds. This is not enough. The definition of task elements for research and development should be overriding. It is the essential element of technology base management since you are positioning your company for the future.

Positioning your company properly for the long term is not an easy process; in fact, there doesn't seem to be a unanimous opinion as to how one should address the problem. Serious study of the mood in Washington may be an excellent place to start.

For years, learned industry spokesmen have urged consideration of a whole range of modifications for the systems and procedures employed by government agencies in the procurement of goods and services. Thus, each time we see a restructuring of the management sectors within the executive branch of government, we also see new sets of initiatives intended to streamline and improve effectiveness of whatever system or procedure they address. But, the change process is terribly slow. Frustrations within industry drive competent competitors toward other pursuits—the exact opposite reaction anticipated by the proponents of change.

The mood of Congress, best reflected during the annual and overly complex federal budgeting process, is a critical element to consider in plotting the future of your company. Are your programs—that is, your target programs—axed, reduced in scope, stretched in time, or allowed to continue for another year? Can you plan the future of your company? Can you position your company properly? Probably not! One very successful marketing manager has told me: "I try to read and interpret the signs. I formulate the best possible plan. I save space for change. I must be flexible."

Even today, the very existence of a long listing of initiatives attests to the belief held by most industry and government leaders that changes in method and procedure are needed. However, there appears to be no clear-cut path to change; the obstacles are formidable. The primary obstacle is the inertia of the system or systems where change is contemplated. You can't stop a locomotive by placing your foot on the track.

Change for the sake of change is still a poor excuse for incorporating modifications to any system, regardless of its complexity. When initiatives address core issues, the best one can hope for is minor mutually beneficial but peripheral modifications.

Because of the flow-down of requirements through prime contractors to almost all lower levels, all marketing managers should continually study the government's changing acquisition policy. To remain inflexible is to invite failure. President Reagan's Executive Order (12352, dated March 1982) addresses the need to achieve a single, simplified Federal Acquisition Regulation. The proposals for change, already documented, will have serious implications for all levels of contract performance within industry. Consequently, to properly position your company for the long haul, you also have to study these implications and interpret them for their impact on your current business plan.

For business organizations dependent upon the export market for a percentage of their annual sales, and for their growth projections, a more serious study of the shape, form and technique of this business sector may be essential. The best example of the changing export market and its impact on American industry is seen in the commercial aircraft sector. The success of Airbus Industries has eroded U.S. aircraft export sales substantially in recent years. Subsidized heavily by the governments of the countries involved, the consortium provides a quality product which is well within the competitive window. These foreign governments also perform the marketing role, alongside their industry counterparts, of promoting wide-body jet aircraft sales all over the world. But, it doesn't stop there. This condition has permeated many other industries, products, and technologies.

Importers of our high technology products and know-how now demand that, as a condition of the sale, we also release the capability to reproduce whatever we sell them. They ask for and are receiving coproduction authority, and even approval and assistance to produce the product for their own use—and export, if they choose. International market shares for U.S. industries are declining. If your company is an exporter, it is now more essential than ever to examine the international marketplace thoroughly. This must be done before you commit your company to any long-term plan based on continued and even expanded sales levels to existing and projected foreign customers. The bottom line is that U.S. industry is competing against foreign governments for market share. The cards are stacked; the odds are unfavorable.

Any long-time observer of marketing managers will confirm that the single most important objective for them is booking the order. Having tasted the rewards of acquiring long-sought contract awards, and thus meeting annual budget goals, the marketing manager and his group usually become less and less concerned with the long-term aspects of their assignment. This may be nothing more than a reflection of the attitudes of a company's top management. The bottom line—the achievement of the financial objectives for the current period—is the yardstick by which they are measured. Seldom are they judged on how well they are structuring their company for the future. The pressures are too great to show current-year profitability. Lower ratios caused by an investment in the future are intolerable, unless management can produce solid evidence that these investments will provide dramatic improvements in growth and profitability.

This is the real challenge for high technology marketing personnel—*developing the case for the future*. And, their investigative work must be introspective as much as it is reflective of the marketplace environment. The achievement of cost and technical competitiveness is derived from within the company. Acceptance in the marketplace is based on performance against current job requirements. The customer is much less concerned about a company's profitability than he is about the delivery of quality products, on schedule, and in full conformance with his specifications. This, then, is the struggle for balance that marketers must pursue.

INDEX

Acquisition:
  assets, 98
  covert (of data), 182
  government system, 134
  of market data, 165
  new business, 40, 54, 161
  policy, 210
  program, 34, 61, 131
Advertising:
  agency, 208
  to aid win, 105
  concept for, 199
  continuity of, 199
  to create image, 199
  to influence, 204
  planning, 200
  publication selection, 202
  timing of, 200
Affordability, of product, 95
Agency, government, 47
Analysis, of relationships in market, 83
Assessments:
  plan implementation, 17
  of risk, 106
Associations (Industry):
  American Defense Preparedness Association, 149
  Electronic Industries Association, 112, 149
  National Contract Management Association, 136, 149
  National Security Industrial Association, 149
  Technical Marketing Society of America, 135
Audit, for contract award, 146
Aviation Week & Space Technology, 198
Award fee, see Contracts

Best and final, see Proposals
Bid:
  decision, 102
  funding, 102, 105
Bidders:
  briefings, 75
  list, 77
B & P, see Funding
Budget, budgeting:
  basis for, 157
  control of, 151
  flexibility of, 157
  general considerations, 55, 97
  zero-based, 156, 202
Business:
  acquisition assets, 98
  acquisition plan, 161
  budgeted, 102
  capture of, 102
  core, 96
  cyclic, 131
  diminishing, 14
  emerging, 14

213

Business *(Continued)*
  environment, 23, 31
  ethics, 182
  evaluation of, 99
  expansion, 96
  opportunities, 30, 50
  risk, 127
  should be, 12
  social responsibility of, 24
  sustaining, 14, 96
  targets, 22, 67
  what it is, 12
  will be, 12
Business Development:
  definition, 25, 61, 103
  organization, 7
Business Plan:
  elements of, 25
  related documentation, 160
Business planning:
  departure from, 16
  ITT, 15
  and marketing communications, 197
  see also Planning
Buy America, 77
Buyer, as adversary, 148
Buy-in, 142

Capture, *see* Business
Cash cow, 61, 96
Classified:
  data, 168
  procurements, 71
Clearances, *see* Security
Commerce Clearing House, 135
Communications:
  by management, 190
  marketing, *see* Marketing
Company:
  environment (business), 10
  mission, 11, 12
  objectives, 11, 12
  other divisions of, 43
  perception of, 194
  performance, 100
  position, positioning, 26, 208
  proprietary data, 180
  social responsibility, 7
  strategy, strategies, 12, 73

Compensation, *see* Marketer
Competitive, 34, 56
Competitor:
  analysis, 73, 77
  as customer, 78
  data base for, 73
  direct, 78
  dossier for, 74
  environment (business), 10
  foreign, 76
  government, 76
  identification, 74
  indirect, 77
  proprietary data, 180
  relation to, 34
  slotting, 76
Computer-aided Design, 6
Consultants, 187
Consumer products, 82
Contract Administration, 52
Contract Management, 136
Contractor, defense, 152
Contracts:
  cost, 139
  fixed price, 137
  with government, 69
  incentive, 138
  multiyear, 139
  negotiations, 135
  pricing for, 140
  redetermination of, 138
  study, 104
  terms and conditions of, 145
  types of, 133, 136
Control:
  of budgets, 152, 155
  by customer, 40
  data library, 162
  by people, 159
  for proposals, 118
Core:
  business, 96, 99
  capability, 76
Cost:
  allocable, 147
  allowable, 147
  effect of risk, 128
  estimating, 125, 130
  overrun, 54, 141

INDEX

performance, 100
proposal, 120, 122
reimbursement of, 139
sharing, 138
start-up, 140
Criteria, for bid evaluation, 103
Curve-fitting, for technology, 93
Customer:
  analysis, 30
  as competitor, 78
  data base for, 65, 68
  debriefings by, 102
  dedicated to, 40
  developing, 66
  development of, 63
  environment (business), 10
  in industry, 70
  partner with, 63
  proprietary data, 181
  relations, 42
  source selection, 106, 125

Data:
  analysis, 24
  for business planning, 18, 166
  for competitor, 73
  library control of, 167
  service organizations, 85
  sources of, 167
Debriefing, by customer, 102, 125
Defense Acquisition Regulation, 135, 136, 145
Defense Contract Administration Services, 135
Department of Defense, 5, 47, 66, 75, 76, 84, 86, 95, 105, 112, 182, 202
Department of Transportation, 8
Deployment, 60
Design to Cost, 124
Design to Unit Production Cost, 124
Development:
  advanced, 105
  cycles, 59
  engineering, 105
  full scale, 59, 60, 105
  product, 59
Dialogue, with customer, 94
DMS, Inc., 85
Drucker, Peter F., 17, 189

Economy:
  conditions of, 110
  price adjustment for, 138
Electronic Industries Association, 112
Emphasis, shifting, 17
Employee:
  from government service, 187
  motivation, 189
Engineering, marketer candidates, 187
Engineering Development, program for, 33
Estimating, for proposals, 122, 141
Evaluation, of proposals, 125
Exhibits, 193, 198

Federal Acquisition Regulation, 136, 210
Finance, function, 36
Fixed Price, see Contracts
Flexibility, in planning, 19, 189
Forecasting:
  for long range, 113
  visibility, 21
  windows, 109
Freedom of Information Act, 75, 170
Free enterprise, 23
Frost & Sullivan, Inc., 85
Funding:
  bid and proposal, 102, 105, 151
  discretionary, 152
  by government, 57, 105, 110, 152
  research and development, 151

Gansler, Jacques S., 6
General Services Administration, 70, 169
Government:
  budgets, 123
  classified information, 179
  foreign policy, 110
  laboratories, 71
  procurement, 67
  procurement system, 69, 133
  project manager, 33
  spending, 51
Government Contracts Reporter, 135
Government offices, see specific departments and offices
Growth:
  business, 13, 34
  projection, 21

INDEX

High technology:
  definition, 3
  marketing of, 3
  transition of, 53
  see also Technology
Honorarium, 197

Incentive, see Contracts
Independent Research and Development, see Research and Development
Industrial Security Manual, 183
Inflation, as market factor, 86
Information:
  classified, 179
  compartmented, 184
  management of, 174
  into priorities, 175
  proprietary, 74
  segregation of, 176
Initiatives:
  for change, 209
  for investment, 61
Integrated circuits, 6
Intelligence, gathering of, 74
Interfaces, within system, 30, 92
Interoperability, within system, 93
Insertion, of technology, 60
Investment:
  capital, 86
  initiatives, 61
  for lowest price, 56
  by small business, 153
  strategic, 23
  technology base, 56, 57
Invitation for Bids (IFB), 148
Implementation, of strategies, 17, 101
Importers, 210

Joint Venture, 45, 78, 181

Laboratories:
  general, 57
  of government, 71, 76
Lead time, 60, 100, 106, 119
Liar's Poker, 144
Library (Data), 167, 169
Licensing, 115, 131
Long-range:
  business opportunity, 50, 131
  forecasting, 113

Management:
  in business acquisition, 42
  communications, 190
  cost audit, 146
  failure effect, 132
  of failures, 22
  of information, 174
  by objectives, 189
  participative, 190
  of proposals, 118
  of risk, 129
  of technology base, 61, 208
Manager, for proposals, 119
Margin, setting of, 124
Marketer:
  compensation for, 8
  complacency, 51
  experience of, 187
  social, 188
Marketing:
  active role, 9
  business development, 61
  communications, 72
  consultants, 187
  definition, 7
  government, 5
  passive role, 10
  plan, 162
  post-award, 126
  shared customer, 49
  shared responsibility, 40, 44
  staff planning, 185
  system, 9
  systems analysis, 27
  tactics, 78
  technical, 8
Marketplace:
  analysis, 81
  fragility, 110
  government, 4, 5
  and high technology, 53
  position in, 25
  representation in, 48
Markets:
  analysis (analysts), 84
  available, 87
  intelligence, 74
  perceived and real, 86
  research, 82
  served, 87

share of, 88
sizing, 84
studies, 85, 168
total, 82
Military spending, 5
Mission:
  for company, 11
  focus, 188
Motivation, 189
Multiyear, see Contracts

NASA, 8, 136
Nation's Business, 6
National Contract Management Association, 136, 149
National Security Industrial Association, 179
Need:
  early warning, 95
  expression, 93
  fulfillment, 93, 101
  marketplace, 24, 94
  technology created, 93
Negotiation:
  contract terms, 45, 140, 146
  practices, 134

Objectives:
  contract pricing, 134
  definition, 12
  financial, 36
  focus on, 37
Obsolescence, of product, 13, 88
Office of Federal Procurement Policy, 70, 136
Office of the Federal Register, 169
Office of Management and Budget, 70, 136, 156
Organization:
  company, 43
  relationships, 42
Overhead, of company, 86, 151

Papers and Presentations, 196
Paperwork, 160
Perception, of company, 194
Performance:
  customer evaluation of, 43
  shortfall, 44

Personnel:
  selection of, 186
  utilization for planning, 15
Planning:
  for advertising, 200
  business, 11, 13, 14, 15, 197
  business acquisition, 160
  definition, 12
  flexibility, 19, 163
  investment, 22, 115
  long-range, 16, 19, 113
  measurement, 115
  realism, 86
  short-range, 16
  strategic, 2, 11
  weakness, 103
Positioning, 208. See also Company
Price-Pricing:
  buy-in, 142
  levels of, 56, 76
Prime contractor, 30, 33, 70
Procurement:
  classified, 71
  cycles, 110
  four-step, 148
  government system, 134
  practices of, 48, 133
  sole-source, 72
  two-step, 148
Products:
  discarding of, 14
  improvement of, 60
  literature, 193
  obsolescence, 13, 88
  replacement, 14
Product lines, protection of, 98
Professional organizations, see Associations
Profitability, 13, 34
Program Evaluation Research Technique (PERT), 112
Programs:
  budgeted, 67
  key, 98
  with no leverage, 98
  opportunities, 97
  review process, 130
  for risk reduction, 130
Proposals:
  balanced attack, 126
  best and final, 143

Proposals (Continued)
  content of, 121
  cost, 33, 120
  evaluation of, 19, 125
  leadership for, 118
  management, 106
  preparation of, 36, 118, 120
  pricing, 182
  ranking of, 119
  solicited, 117
  technical, 33, 58, 106, 120
  unsolicited, 117
Proprietary (data), 163, 179
Public relations, 193

Request for Proposal, 102, 148
Request for Quotation, 129, 148
Research, market, 82
Research and development:
  allocations for, 58
  company investment, 61, 105, 208
  contribution of, 152
  fixed price contract for, 137
  funding for, 151
  general, 33, 34, 35
  government investment, 57
  planning for, 113
  report of, 152
  seed money for, 100, 121, 153
  structuring, 153
  universities, programs for, 76
Responsibility, for product line, 51
Risk:
  assessment study, 106, 138
  definition, 128
  evaluation of, 45
  factors, 22
  identification of, 56, 129
  reduction of, 129
  in target selection, 96
  technical, 102
Robotics, 6

Sales:
  margin, 36
  organization, 7
Sanitized, 75
Schedules, scheduling:
  effect of risk, 128
  elapsed time, 111

  event cycles, 112
  event windows, 112, 113
  flow charting, 111
  by government, 111
  optimistic, 112
  PERT, 112
  pessimistic, 112
  relationships, 54
Security:
  of classified data, 169, 170
  clearances, 169, 182
  company plan for, 163
  levels of, 183
  for procurements, 71
Selling, responsibility for, 40
Small business:
  feeder suppliers, 64
  organization, 4
  and R & D, 153
Social programs, 24
Sole source, see Procurement
Source selection, 106, 125
Strategic planning, see Planning
Strategy, strategies:
  best and final offer, 144
  for company, 73
  definition, 12
  implementation, 101, 107, 142
  for investment, 22, 56
  tailoring of, 92
  for winning, 75, 78, 91, 101, 107
Subcontractor, 30
Superintendent of Documents, 112
Suppliers:
  relationship to, 45
  teaming with, 46
Synergism, 45, 78
Systems:
  acquisition, 60
  analysis concepts, 29
  concept for, 27
  diagram for, 32
  interactive within, 30
  interface, 30
  management information, 174
  marketing, 9
  numerical assignment to, 36
  objectives for, 27
  paperwork, 160
  subsystems of, 34

INDEX

Tailoring, of win strategy, 92
Target, targets:
  budgeted, 104
  complimentary, 56
  conservatism about, 99
  disappearance of, 131
  evaluation of, 104
  new business, 54
  programs, 55, 56, 67
  selection of, 91, 96, 102
Team, teaming, 45, 78, 180
Technical:
  issues, 46
  papers and presentations, 196
  proposal, 120
Technical Marketing Society of America, 135
Technology:
  accessible, 54
  acquisition, 114, 131
  assessment of, 53
  base, 9, 31, 35, 82, 152
  base assessment, 61
  base deficiencies, 55, 57, 100
  base limitations, 54
  base management, 22, 208
  breakthrough, 20
  condition of, 17, 56
  freeze, 59
  insertion, 60
  investment in, 57
  leading edge of, 54
  matching to market, 70
  maturity, 60
  obsolescence, 60
  premature application of, 128
  product content, 89
  see also High Technology
Termination, 44
Test and evaluation, 33
Threat, to national security, 110
Timing, for forecasting, 110
Trade associations, see Associations
Trade shows, 193, 198
Trends:
  course setting, 207
  evaluation of, 130

United States Government Manual, 169
U.S. Government Printing Office, 112, 135, 169
User, customer organization, 47, 68, 92

VHSIC, 57
Vulnerability, of market, 85

Win:
  can, cannot, 103
  strategy for, 75, 91, 101, 103
Windows:
  for cost structuring, 123
  event scheduling, 112, 113
Work packages, 17, 26

Zero-base budgeting, 156, 202